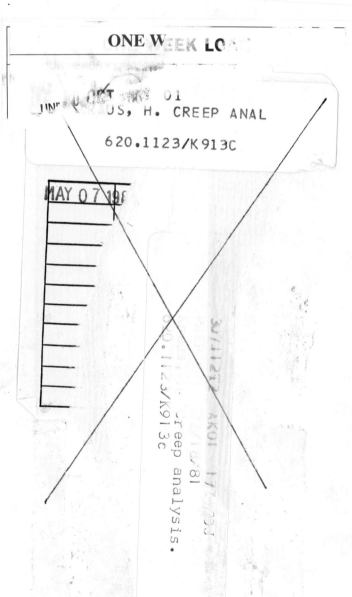

CREEP ANALYSIS

Creep Analysis

HARRY KRAUS

Professor and Director of Engineering Studies
The Hartford Graduate Center
Hartford, Connecticut

A WILEY-INTERSCIENCE PUBLICATION

JOHN WILEY & SONS
New York • Chichester • Brisbane • Toronto

Copyright © 1980 by John Wiley & Sons, Inc.

All rights reserved. Published simultaneously in Canada.

Reproduction or translation of any part of this work beyond that permitted by Sections 107 or 108 of the 1976 United States Copyright Act without the permission of the copyright owner is unlawful. Requests for permission or further information should be addressed to the Permissions Department, John Wiley & Sons, Inc.

Library of Congress Cataloging in Publication Data

Kraus, Harry.
   Creep analysis.

   "A Wiley-Interscience publication."
   Includes index.
   1. Materials—Creep. I. Title.

TA418.22.K7   620.1'1233   80-15242
ISBN 0-471-06255-3

Printed in the United States of America

10 9 8 7 6 5 4 3 2 1

To my wife Jane

To my Jane

PREFACE

Many of today's graduates are likely to find themselves employed in the design and construction of nuclear or fossil power generating equipment, aircraft gas turbine propulsion engines, refinery equipment, or process equipment. In order to achieve greater operating efficiencies in all of these applications, operating temperatures are being pushed to higher and higher levels. Those performing stress analysis of such equipment are now dealing, and will continue to deal, with the creep problem: the time dependent deformation and/or failure of components and structures.

The purpose of this book is to introduce the reader to the analysis of the creep problem. The coverage of subjects and the point of view are selected to present a combination of the foundations and the applications of creep analysis so that the reader may gain an understanding and appreciation of each of these aspects. The book is aimed primarily at graduate students at the intermediate level in mechanics and mechanical engineering specializing in solid mechanics. Stress analysts in industry will find the book a useful introduction that will equip them to read further into the literature of solutions to technically important creep problems. Research specialists will find it useful as an introduction to current analytical work in creep. This volume is not intended to be an exhaustive treatise on creep but, rather, a broad introduction from which each reader can pursue his or her interest further. The book is largely an outgrowth of classroom notes I have used in teaching courses on creep in both degree and nondegree programs at the Hartford Graduate Center, which is affiliated with the Rensselaer Polytechnic Institute. These notes were also used for in-house courses on creep that I have conducted at the Exxon Research and

Engineering Company, the Pratt & Whitney Aircraft Group of the United Technologies Corporation, and the Sandia Laboratories.

To combine the fundamentals and the applications of the theory of creep, the book is divided into nine chapters. Chapter 1 presents creep as a class of mechanical behavior that can be observed in the laboratory. Chapter 2 develops mathematical models of uniaxial and multiaxial creep behavior. Chapter 3 presents solutions to creep deformation and stress relaxation problems, while Chapter 4 treats these with approximate analytical techniques. Chapters 5, 6, and 7 present material on creep rupture, creep buckling, and creep ratchetting, respectively. Then, in Chapter 8, numerical analysis of various aspects of the creep problem is covered. Finally, Chapter 9 brings in the influence of creep on the fatigue problem. The material on creep buckling, approximate analytical solutions, numerical solutions, and creep fatigue interaction has not appeared previously in book form. It represents the results of contemporary work in the field.

By way of mathematical preparation the book assumes that the reader has had courses in advanced calculus and, if possible, numerical methods. In addition it is assumed that the reader has had courses in advanced strength of materials and elasticity. A course in plasticity would be beneficial as well, but is not necessary.

It is felt that the material can be covered in a one semester course. More than enough material is provided in each chapter for the instructor to cover what is of greatest interest to the class and assign the rest for reading. If desired, it would be possible to omit either Chapters 8 or 9 from a one semester course. If the instructor and class members are oriented toward problem solving, then Chapter 8 on numerical analysis should be included. If, on the other hand, the class is oriented toward mechanical behavior, then Chapter 9 should be included. If further shortening is required, I would suggest that Chapter 4 on approximate methods can be deleted.

The book represents the culmination of knowledge I have gained during the last ten years. During this time I have taught, done research, and consulted on the creep problem. I am particularly grateful to my colleagues on the Subcommittee on Elevated Temperature Design of the Pressure Vessel Research Committee (an arm of the Welding Research Council) for many stimulating discussions over the years. The Committee's support, via grants, of my work in the field is also acknowledged. The chapters on rupture, buckling, ratchetting, and numerical analysis were strongly influenced by the writings of Odqvist, Chern, Burgreen, and Zienkiewicz, respectively. I am indebted to these people.

The text was typed by Judith Rohan, whose capable, patient efforts are hereby acknowledged.

Harry Kraus

Hartford, Connecticut
August 1980

ACKNOWLEDGMENTS

Figures 1.12, 1.13, 1.14, 2.3, 2.4, 2.5, 2.7, 7.9, 7.11, 8.10, 8.14, 8.15, 8.16, 8.21, 8.22, 8.23, 8.24, 9.1, 9.11 and 9.12 are reprinted by permission of the Oak Ridge National Laboratory, operated by the Union Carbide Corporation under contract W-7405-eng-26 with the U.S. Department of Energy.

Figures 1.11, 6.3, 8.2, 8.4, 8.5, 8.6, 8.7, 8.8, 8.9, 8.11, 8.12, 9.2, 9.3, 9.4, 9.6, 9.7, 9.8, 9.9, and 9.10 are reprinted by permission of the American Society of Mechanical Engineers.

Figures 2.1, 2.2, 3.1 and 6.1 are reprinted by permission of the McGraw-Hill Book Company, Inc.

Figures 7.1, 7.2, 7.3, 7.4, 7.5 and 7.7 are reprinted by permission of the C.P. Press.

Section 8.4g and Figures 8.17, 8.18, 8.19 and 8.20 are presented by permission of the Pratt & Whitney Aircraft Group of the United Technologies Corporation.

CONTENTS

1 **EXPERIMENTAL OBSERVATIONS OF THE MECHANICAL BEHAVIOR OF STRUCTURAL METALS**   1

   1.1 Introduction, 1
   1.2 The Tensile Test, 2
   1.3 Instantaneous Behavior: Elasticity and Plasticity, 3
   1.4 Time Dependent Behavior: Creep, Recovery and Relaxation, 6
   1.5 Cyclic Loadings: Fatigue at Room and Elevated Temperatures, 11
   1.6 Closure: Failure of Structures, 16
   References for General Reading, 17

2 **MATHEMATICAL MODELING OF CREEP PHENOMENA**   18

   2.1 Introduction, 18
   2.2 Uniaxial Creep Models, 18
   2.3 Multiaxial Creep Models, 27
   2.4 Formulation of the Creep Problem, 32
   2.5 Relaxation, 35
   2.6 Steady Creep Approximation—Elastic Analogy, 36
   2.7 Closure, 39

APPENDIX: Procedures for Stress Reversals in Multiaxial Creep, 39

References, 44

General References for Further Reading, 45

Exercises, 45

3 CREEP DEFORMATION, STRESS RELAXATION, AND RECOVERY PROBLEMS   46

3.1 Introduction, 46

3.2 Creep Deformation Problems, 47
- 3.2.1 Creep of a Thin Pressurized Tube with Closed Ends, 47
- 3.2.2 Creep of a Thick Pressurized Tube with Closed Ends, 49
- 3.2.3 A Statically Indeterminate Two Bar Structure, 52

3.3 A Stress Relaxation Problem, 57

3.4 Recovery in a Statically Indeterminate Two Bar Structure, 59

3.5 Closure, 61

References, 62

Exercises, 62

4 APPROXIMATE ANALYTICAL TECHNIQUES   64

4.1 Introduction, 64

4.2 The Reference Stress Method, 64
- 4.2.1 Development of the Reference Stress Method for Creep under Steady Loads, 65
- 4.2.2 Extension to Combined Loadings, 75
- 4.2.3 Extension to Thermal Loadings, 77
- 4.2.4 A Reference Stress Method for Stress Relaxation, 79
- 4.2.5 The Skeletal Point Concept, 82
- 4.2.6 Experimental Verification of the Reference Stress Method for Steady Loads, 82

4.2.7 Application to Creep Deformation Under Variable Loads, 86
4.2.8 Conclusions on the Reference Stress Method, 90

4.3 Bounding Techniques Based on Virtual Work, 90
4.3.1 The Principle of Virtual Work, 90
4.3.2 Application to Creep Problems, 92

4.4 Closure, 95

References, 95

5 CREEP RUPTURE 98

5.1 Introduction, 98

5.2 Creep Rupture Data, 99
5.2.1 Larson-Miller Parameter, 100
5.2.2 Manson-Haferd Parameter, 103

5.3 Analysis of the Rupture Process, 104
5.3.1 The Damage Concept, 104
5.3.2 Multiple Stress Levels, 106

5.4 Analysis of the Creep Rupture of Structures, 108
5.4.1 Creep Rupture of a Tensile Specimen, 109
5.4.2 Extension to Multiaxial and Nonhomogeneous Stress States, 114

5.5 Approximate Analysis of the Creep Rupture of Structures with the Reference Stress Method, 118

References, 123

Exercises, 124

6 CREEP BUCKLING 126

6.1 Introduction, 126

6.2 Creep Buckling of a Column, 127
6.2.1 Analytical Solution, 127
6.2.2 Tangent Modulus Approach, 130

6.3 Creep Buckling of Shell Structures, 134
6.3.1 The Reference Stress Method, 134

6.3.2 The Critical Effective Creep Strain Method, 136
6.3.3 Comparison of Methods, 141

6.4 Closure, 147

References, 148

Exercises, 148

7 CREEP RATCHETTING 150

7.1 Introduction, 150

7.2 A Thin Tube Subjected to Constant Internal Pressure and Cyclic Thermal Stresses, 151
7.2.1 The Elasto-Plastic Problem, 153
7.2.2 Effect of Stress Relaxation due to Creep, 167
7.2.3 A Bounding Technique for Strains due to Creep Ratchetting, 170

7.3 Creep Ratchetting of a Tube as Observed in the Laboratory, 172

7.4 Closure, 174

References, 176

8 NUMERICAL ANALYSIS OF CREEP 177

8.1 Introduction, 177

8.2 Summary of the Finite Element Method, 178

8.3 Application to Creep Analysis, 185
8.3.1 Method of Initial Strains, 186
8.3.2 Extraction of Time for Strain Hardening Creep, 187
8.3.3 Computer Programs for Creep Analysis, 189

8.4 Illustrative Solutions, 192
8.4.1 Stress Redistribution in a Thick Pressure Vessel with an Ellipsoidal Head, 192
8.4.2 Stress Redistribution in a Rotating Disc, 195
8.4.3 Creep Ratchetting of a Thick Tube, 198
8.4.4 Creep of a Straight Tube Under Combined Bending, Pressure, and Thermal Loads, 201
8.4.5 Creep and Relaxation of a Circular Plate Under Prescribed Transverse Deflections, 203

    8.4.6 Creep Buckling of an Axially Compressed Cylindrical Shell, 205
    8.4.7 Creep of a Rotating Gas Turbine Seal Ring, 207
    8.4.8 Cyclic Creep of a Primary Closure Seal for a Nuclear Power Plant Component, 211

  8.5 Closure, 212

  References, 215

9 CREEP-FATIGUE INTERACTION 218

  9.1 Introduction, 218

  9.2 A.S.M.E. Code Procedure, 219

  9.3 Strain Range Partitioning, 226

  9.4 Frequency Separation, 234

  9.5 Closure, 239

  References, 241

  AUTHOR INDEX 243

  SUBJECT INDEX 247

CREEP ANALYSIS

CHAPTER 1

EXPERIMENTAL OBSERVATIONS OF THE MECHANICAL BEHAVIOR OF STRUCTURAL METALS

1.1 INTRODUCTION

Creep involves the time dependent deformation and fracture of materials. The phenomenon is accelerated by an increase in the stress or the temperature. It can be described as being the most general type of material behavior. In this book we devote our attention to the modeling and analysis of the creep phenomenon in metals so that it can be incorporated into the design of structures operating at elevated temperature. In this chapter we review various aspects of the mechanical behavior of structural metals. This is intended to introduce the creep phenomenon and to place it within the context of the mechanical behavior of materials. This includes discussions of both deformation and failure modes and the manner in which these fit into the design analysis process. Our approach is phenomenological; we leave the explanation of the phenomena to texts on mechanical behavior (see References).

We first discuss the tensile test, which is used to gather all of our information on behavior. This is followed by a discussion of instantaneous behavior under steady loads, namely, elasticity and plasticity. Then we take up time dependent behavior, that is, creep, relaxation, and recovery.

After that we consider the effect of cyclic loads and introduce the fatigue problem both at room temperature and at elevated temperature. The chapter closes with a discussion of failure modes that the design analyst must take into account to produce a safe structure.

1.2 THE TENSILE TEST

The basic vehicle for bringing out the behavior of structural materials is the uniaxial tensile test. A cylindrical specimen of the material is made and placed in a tensile testing machine. A central gage length of length l_0 and area A_0 is defined and then the load P is applied. As the load is increased, the length increases and two quantities are calculated. The nominal stress is

$$\sigma_n = \frac{P}{A_0}$$

and the conventional (engineering) strain measured is

$$\varepsilon = \frac{l - l_0}{l_0}$$

where P and l are the current load and length, respectively. If the deformations become large and if, as is usually the case, necking down occurs, it is also possible to define a true stress

$$\sigma = \frac{P}{A}$$

where P and A are the current load and area, respectively. Similarly, a natural strain can be defined as

$$d\bar{\varepsilon} = \frac{dl}{l}$$

where dl is an increment in the length and l is the current length. Integration then gives

$$\bar{\varepsilon} = \int_{l_0}^{l} \frac{dl}{l} = \ln \frac{l}{l_0}$$

and furthermore, since $\varepsilon = l/l_0 - 1$,

$$\bar{\varepsilon} = \ln(1+\varepsilon)$$

$\bar{\varepsilon}$ is also called the logarithmic strain. The true and nominal stresses can be related to each other by imposing the experimentally observed constancy of the volume of the deformation process at high strains. Thus if the constancy of volume is expressed as

$$A_0 l_0 = A l$$

and the true stress is

$$\sigma = \frac{P}{A} = \frac{Pl}{A_0 l_0} = \sigma_n \frac{l}{l_0}$$

then we obtain

$$\sigma = \sigma_n (1+\varepsilon)$$

Thus it is important to be aware of which stress and strain measures are being cited in any report of a tensile test. At small strains it hardly matters, but at large strains it does.

1.3 INSTANTANEOUS BEHAVIOR: ELASTICITY AND PLASTICITY

Now to return to the tensile test we observe several things as the load is increased (see Fig. 1.1). First, most metals demonstrate a steep, straight line relationship between stress and strain up to a limit known as the yield stress. Moreover, if the stress is removed prior to reaching the yield stress, the strain returns to zero. The straight line relationship and the perfect memory of the original stress-free state characterize linear elastic behavior. The slope of the curve is known as the elastic, or Young's modulus, E. Beyond the yield stress σ_y the relationship is no longer linear since progressively less stress is required to produce a given strain change. This continues up to a maximum stress, known as the ultimate stress, σ_u. Beyond this the area begins to drop rapidly. The nominal stress also drops until the specimen fractures at strain ε_f. Now suppose that the specimen has only been loaded to point A. If the load is removed when the stress reaches σ_A two interesting things occur. First, the specimen unloads along a line that is parallel to the initial elastic line; that is, it unloads elastically. Second, the strain does not return to zero since a residual plastic strain

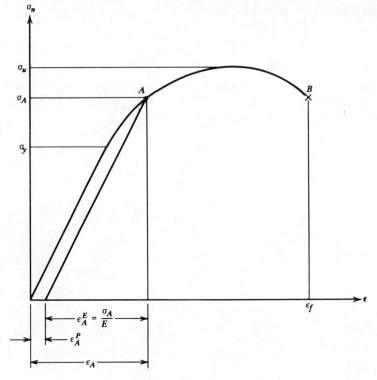

Fig. 1.1 The tensile test.

remains in the specimen. This plastic strain is equal to the total strain achieved at A minus the elastic strain that is recovered, or

$$\varepsilon_A^P = \varepsilon_A - \varepsilon_A^E = \varepsilon_A - \frac{\sigma_A}{E}$$

where the subscript A associates the strain with the stress at point A. Now if the specimen is reloaded, the behavior is linear and retraces the unloading line until the stress reaches σ_A once more. That is to say, now the linear behavior continues beyond the initial point of departure from linearity, σ_y. The material is said to be work hardened. Further increases in load rejoin the curve shown until the specimen fractures at B. The behavior between the yield point and the fracture point, including the unloading and reloading, is characteristic of plasticity.

The strain at fracture is used as a measure of ductility. That is, if the fracture strain is "large" the material is said to be ductile and if it is

Instantaneous Behavior: Elasticity and Plasticity

"small" it is said to be brittle. "Large" and "small" are usually relative to the strain at yield. Ductility is then defined by any of the measures

$$\varepsilon_f = \frac{l_f - l_0}{l_0} \cdot 100\%$$

$$\varepsilon_f = \frac{A_0 - A_f}{A_0} \cdot 100\%$$

$$\bar{\varepsilon}_f = \ln \frac{l_f}{l_0} \cdot 100\%$$

$$\bar{\varepsilon}_f = \ln \frac{A_0}{A_f} \cdot 100\%$$

where f refers to the fracture state. Note that the first two equations are related, as are the last two, through the constant volume observation. Thus ductility is expressed in terms of length changes or area changes.

The stress-strain behavior that we have seen, namely, the initial elasticity followed by plasticity to fracture, does not depend upon time. As soon as the stress changes, the strain changes. Elastic and plastic behavior are thus said to be instantaneous. They do, however, depend on the rate of straining. That is, if the tensile test is performed at different strain rates from very slow to very fast the typical effect on the stress-strain curve is shown in Fig. 1.2. This shows that an increase in strain rate causes an

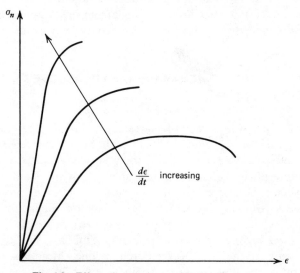

Fig. 1.2 Effect of strain rate on the tensile test.

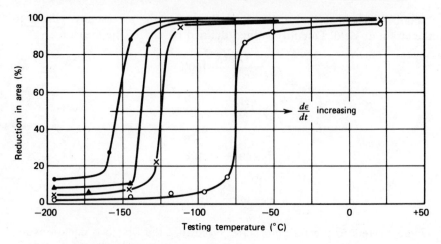

Fig. 1.3 Ductility as a function of temperature and strain rate.

increase in the elastic modulus, yield stress, and fracture stress, and a decrease in the ductility. This is typical of many materials. Changes in temperature also influence the stress-strain diagram. That is, usually an increase in the test temperature decreases both the yield and the elastic modulus. The temperature also influences the fracture ductility, as shown in Fig. 1.3. This shows that there is a transition in behavior as the temperature increases. The drastic change occurs at what is known as the transition temperature. It marks a change from brittle to ductile behavior. The transition temperature is in turn elevated by an increase in strain rate.

1.4 TIME DEPENDENT BEHAVIOR: CREEP, RECOVERY, AND RELAXATION

The most important influence of increasing the temperature is to bring in the time as a factor that must be dealt with. Such time dependence is the chief characteristic of creep. Fig. 1.4 shows a typical creep curve. Now the salient feature of the behavior is that, at constant stress and temperature, strain develops as shown on the curve. At $t=0$ the curve shows an instantaneous response ε_0, which, depending on the magnitude of the stress, could be elastic or elasto-plastic. The creep curve generally arranges itself into three regimes as shown: primary, secondary, and tertiary. Creep rupture occurs at the end of the tertiary zone. Materials differ in the arrangement of the three regimes. Some have hardly any secondary, while

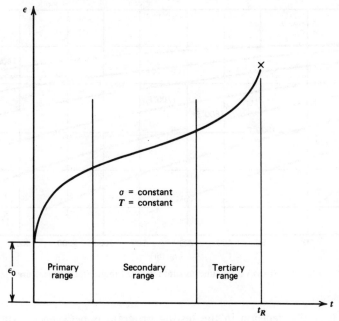

Fig. 1.4 Creep curve.

others have hardly any tertiary creep, and so on. The curve also shows that for every stress rupture eventually occurs.

The relationship of stress and time to rupture is shown in Fig. 1.5. There it is seen that as stress goes down the time to rupture increases. Furthermore, as the temperature increases the time to rupture decreases. Rupture occurs immediately if the fracture stress (see Fig. 1.1) is applied to the specimen. Any stress less than this value takes some time to cause rupture. In examining the rupture curves we should point out that on the left portion the ruptures are observed to be ductile, whereas on the right portion the ruptures are observed to be brittle, using area reduction, say, as a measure of ductility.

The creep curve shown in Fig. 1.4 is strongly influenced by stress and temperature. The influence of stress at constant temperature is shown in Fig. 1.6, while the influence of temperature at constant stress is shown in Fig. 1.7. Note that both show thresholds below which no noticeable creep is observed.

The discussion of true and nominal stress as well as conventional and natural strain pertains to creep as well as it does to elasticity and plasticity.

Experimental Observations of the Mechanical Behavior of Structural Metals

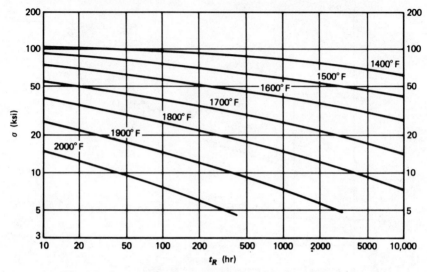

Fig. 1.5 Stress versus time to rupture as a function of temperature.

In fact here the decision in the testing program is between constant load and constant stress. Each gives vastly different results as creep strains develop. Constant load is easy to achieve in the test, whereas constant stress requires complicated feedback devices, which reduce the load as the area reduces to keep the stress constant. Metallurgically creep depends on the stress.

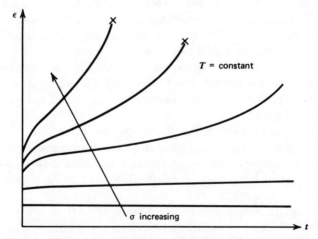

Fig. 1.6 Effect of stress on the creep curve at constant temperature.

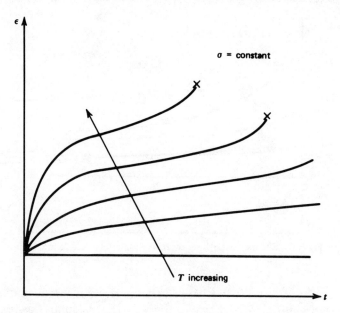

Fig. 1.7 Effect of temperature on the creep curve at constant stress.

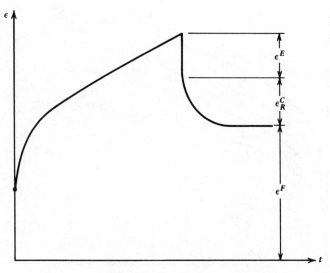

Fig. 1.8 Creep recovery.

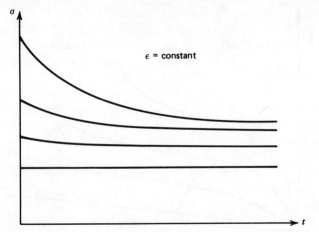

Fig. 1.9 Effect of stress level on relaxation at constant temperature.

Now it is well to ask what happens if the stress is removed during the creep test. The behavior is shown in Fig. 1.8. The instantaneous response is that the elastic strain $\varepsilon^E = \sigma/E$ is recovered. After that there is a certain amount of strain recovery that becomes asymptotic and then no more strain is recovered. ε_R^C is the recovered creep strain. ε^F is the permanent strain that is made up of irrecoverable plastic strain (if any) and

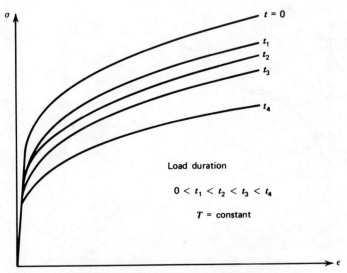

Fig. 1.10 Isochronous stress-strain curves.

creep strain components. Finally, there is one other interesting time dependent phenomenon: stress relaxation. This happens when the strain is held fixed and occurs as shown in Fig. 1.9. Note that as in creep there is a threshold below which no relaxation is observable.

The basic information about time dependent phenomena at elevated temperature is contained in the creep and relaxation curves. Many times it is convenient (as will be shown later in the text) to replot the curves in Fig. 1.6. This is accomplished as follows: At a time t_1 there are several stresses and strains on Fig. 1.6. Plot these pairs on axes of stress versus strain as in Fig. 1.10 and label the curve t_1. At time t_2 there is another collection of stress and strain pairs. Plot these on stress and strain axes as in Fig. 1.10 and label the curve t_2. Proceed for other values of time. The resulting family of curves shown in Fig. 1.10 is known as the isochronous (constant time) stress-strain curves. The curve corresponding to $t=0$ is the instantaneous stress-strain curve at the temperature of the curves.

1.5 CYCLIC LOADINGS: FATIGUE AT ROOM AND ELEVATED TEMPERATURES

Thus far our discussion of elastic, plastic, and creep behavior has been limited to single applications of stress. We have also considered the removal of stress in each instance. For a measure of completeness we must now consider cyclic loads. This brings in two important areas of behavior, namely, fatigue at room temperature and fatigue at elevated temperature. To introduce these briefly for now we start with the room temperature case in which no creep is assumed to take place. If a strain is applied cyclically (on and off, say) it is well known that after a large number of cycles failure eventually occurs. If we plot the strain amplitude versus the number of cycles to failure we get the so-called S-N diagram that is shown in Fig. 1.11. Here the strain amplitude (equal to half the on-off strain range) is multiplied by E to make it a pseudo-stress. Most materials have a stress level, known as the endurance limit, below which the material can stand an essentially infinite number of cycles.

At elevated temperature creep and relaxation have a serious influence on the fatigue process and we must speak of the interaction between creep and fatigue. This is predominantly an interaction between the deterioration due to the cyclic nature of the load and the fact that creep rupture progresses during any hold times that exist in the loading cycles. Now we must be careful to distinguish between continuous cycles, cycling with hold times during which creep occurs, and cycling with hold times during which relaxation occurs. The continuous cycle loading is shown in Fig. 1.12. The

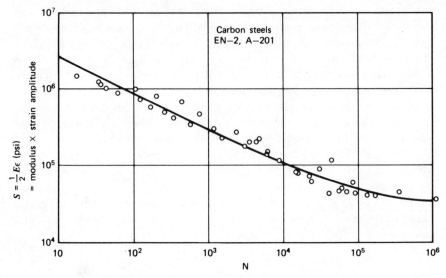

Fig. 1.11 A typical fatigue curve. Courtesy of the American Society of Mechanical Engineers.

strain is controlled as shown in Fig. 1.12(*a*) and the stress response is as shown in Fig. 1.12(*b*). The influence on the fatigue curve is shown in Fig. 1.12(*c*). There it is seen that as the strain rate is increased there are more cycles to failure for a given strain range. This is because, during the slow strain rates, more time is spent near the peak of the stress cycle and creep damage occurs. The cyclic relaxation loading is shown in Fig. 1.13. The strain is controlled as in Figure 1.13(*a*). Now there is a hold time t_h during which the strain is held constant and the stress relaxes. The resulting stress is shown in Fig. 1.13(*b*) and typical test data are shown in Fig. 1.13(*c*). It is seen that a hold time in the tensile part of the cycle reduces the cyclic life and there seems to be a saturation point beyond which not much change occurs, that is, the curves are closer and closer together as hold time increases. The strain slope $\dot{\varepsilon}$ has little effect on this behavior because the time spent at the peak strain is more deleterious than the time spent getting there. A hold in compression could also be introduced. It has been found that for some metals this has no effect on life. Many combinations are possible. The cyclic creep loading is shown in Fig. 1.14. Now it is the stress that is controlled as in Fig. 1.14(*b*) and the resulting strain is as shown in Fig. 1.14(*a*). A tensile hold time on the stress is incorporated into the cycle. The typical fatigue data are shown in Fig. 1.14(*c*). Now the hold time again reduces the cyclic life but there is no saturation as with the cyclic relaxation loading. At this point it should be recognized that these

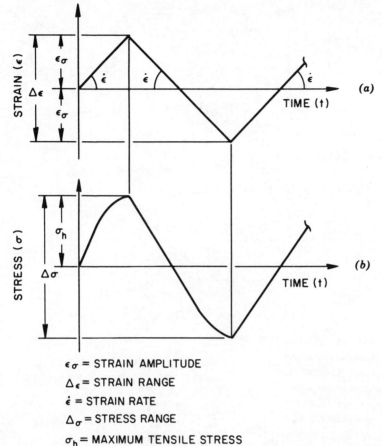

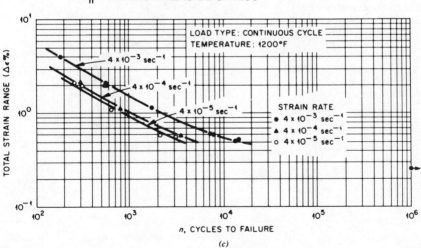

ϵ_σ = STRAIN AMPLITUDE
Δ_ϵ = STRAIN RANGE
$\dot{\epsilon}$ = STRAIN RATE
Δ_σ = STRESS RANGE
σ_h = MAXIMUM TENSILE STRESS

Fig. 1.12 Continuous cycle fatigue data for 304 stainless steel at 1200°F. Courtesy of Oak Ridge National Laboratory.

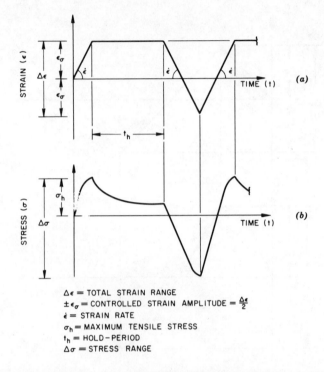

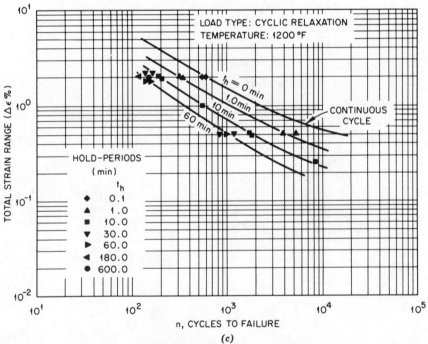

Fig. 1.13 Cyclic relaxation data for 304 stainless steel at 1200°F. Courtesy of Oak Ridge National Laboratory.

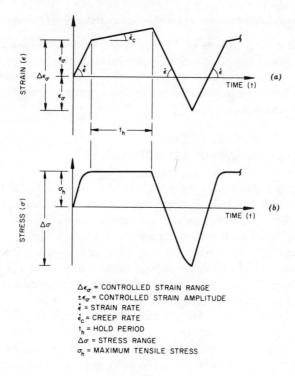

$\Delta\epsilon_\sigma$ = CONTROLLED STRAIN RANGE
$\pm\epsilon_\sigma$ = CONTROLLED STRAIN AMPLITUDE
$\dot{\epsilon}$ = STRAIN RATE
$\dot{\epsilon}_c$ = CREEP RATE
t_h = HOLD PERIOD
$\Delta\sigma$ = STRESS RANGE
σ_h = MAXIMUM TENSILE STRESS

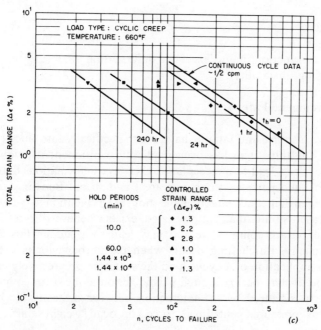

Fig. 1.14 Cyclic creep data for low carbon, high manganese steel. Courtesy of Oak Ridge National Laboratory.

three cycles are idealizations of what happens in the real case of a power plant, say, undergoing a completely general duty cycle. This may be a combination of the three that were given with various hold times in both tension and compression. This makes for an extremely complicated problem that we take up later.

1.6 CLOSURE: FAILURE OF STRUCTURES

In the foregoing discussion we have given a brief introduction to the salient features of the behavior of structural metals at and above room temperature. We have used the uniaxial tensile test as the vehicle for this. We have considered both instantaneous and time dependent behavior and both steady and cyclic loads. The aim has been to place creep and its associated phenomena of relaxation and rupture within the context of the general behavior of materials. The treatment is not intended to be exhaustive; rather, it is meant to point out important areas of behavior. Detailed discussions of each of these are presented in the remainder of the book.

The end purpose of understanding all such behavior is to incorporate it into the design process. This is concerned with the prevention of several modes of failure. Some of these have been explicitly mentioned in the preceding discussion while the remainder have not. They are:

1. Rupture.
2. Fatigue.
3. Loss of function due to excessive deformation.
4. Buckling (instability).
5. Ratchetting.
6. Environmental attack.

The first two, rupture and fatigue, have been specifically introduced in this discussion. The third item has been implicitly mentioned; it involves the fact that loads can lead to unacceptable, excessive deformations that cause parts to rub, fluid passages to change area, and so on. Buckling, ratchetting, and environmental effects have not been mentioned previously. Buckling is a possibility that, simply stated, exists whenever the net stresses across a section are compressive and may lead to a sudden loss in load carrying capacity. Ratchetting occurs when plastic behavior is such that during each application of a cyclic load the structure progressively changes size or shape. It may occur through plastic action alone or in conjunction with creep. It is caused by the interaction between a steady mechanical load such as pressure, centrifugal force, or dead weight and a cyclic mechanical load such as a bending moment or a cyclic thermal load. Environmental attack is caused by the presence of deleterious fluids. Chief among these are chlorides such as sea water and hydrogen compounds such as ammonia, H_2S, and others. Unlike all of the other modes of failure that have been mentioned, environmental attack does not lend itself to mathematical analysis. It must be accounted for, however, and is mentioned here for completeness. Our concern is with the modes that can be analyzed.

All of the modes of failure can occur at room temperature. They are aggravated by elevated temperature and by the fact that this brings in the time dependent aspect of each. Thus the intrepid design analyst must be aware of these modes no matter what the temperature level is. Our particular aim in this book is to contribute to sound elevated temperature design. In pursuit of this we consider only creep and its associated behaviors as they influence the analysis of equipment. Thus we present in various chapters, information that is aimed at an understanding of all of the modes of failure mentioned previously except environmental attack.

REFERENCES FOR GENERAL READING

Finnie, I. and W. R. Heller, *Creep of Engineering Materials*, McGraw-Hill Book Co., Inc., New York (1959).

Garofalo, F., *Fundamentals of Creep and Creep Rupture in Metals*, Macmillan Co., New York (1965).

Hult, J. A. H., *Creep in Engineering Structures*, Blaisdell Publishing Co., Waltham, Mass. (1966).

Johnson, W. and P. B. Mellor, *Engineering Plasticity*, Van Nostrand Reinhold Co., London (1973).

Juvinall, R. C., *Stress, Strain and Strength*, McGraw-Hill Book Co., Inc., New York (1967).

Manson, S. S., *Thermal Stress and Low Cycle Fatigue*, McGraw-Hill Book Co., Inc., New York (1966).

McClintock, F. A. and A. S. Argon, *Mechanical Behavior of Materials*, Addison Wesley Publishing Co., Inc., Reading, Mass. (1966).

Odqvist, F. K. G., *Mathematical Theory of Creep and Creep Rupture*, Clarendon Press, Oxford, (1974).

Penny, R. K. and D. L. Marriott, *Design for Creep*, McGraw-Hill Book Co., Ltd., London (1971).

Suh, N. P. and A. P. L. Turner, *Elements of the Mechanical Behavior of Solids*, McGraw-Hill Book Co., New York (1975).

CHAPTER 2

MATHEMATICAL MODELING OF CREEP PHENOMENA

2.1 INTRODUCTION

In Chapter 1 we have presented the elements of time dependent behavior within the overall framework of the mechanical behavior of materials. To analyze engineering structures that undergo the various phenomena that have been described we need mathematical models of creep and relaxation. In this chapter we develop these models. First, we present uniaxial models that parallel the information discussed in the preceding chapter. Next we extend the development to multiaxial creep and relaxation behavior and present the formulation of a general problem in creep analysis. This is followed by a coverage of relaxation and a presentation of the idea of the steady creep approximation and the associated elastic analogy. An appendix addresses the handling of multiaxial stress reversals.

2.2 UNIAXIAL CREEP MODELS

The creep curve, for example Fig. 1.4, tells us that to model the creep process in one dimension we need an expression of the form

$$\varepsilon^C = F(\sigma, T, t) \tag{2.1}$$

Here ε^C is the creep strain and F is a function of the stress σ, temperature T, and time t. It is customary to assume that the effects are separable. Thus we write

$$\varepsilon^C = \sum_{i=1}^{n} f_i(T)g_i(\sigma)h_i(t) \qquad (2.2)$$

This permits the expression to have several terms.

At this point the development is valid only for constant stress and temperature. For practical problems, however, we require a formulation that is valid for variable stress and temperature. We have two choices for writing such a formulation, that is, we can assume either:

1. The response of the material depends on the present state explicitly.
2. Or the material remembers its past explicitly and responds to the present in a manner that reflects its past history.

The first of these results in what is known as an equation of state formulation while the second gives rise to a so-called memory theory.

Most work in the field is based on the equation of state approach. We select it here on the basis of the following advantages:

1. It has been extensively applied and there is much experience available with it.
2. It is simple to apply and understand.
3. It is easy to incorporate into existing computer programs.
4. It requires a minimum of experimental data: only single step creep tests at various stresses and temperatures.

It has been argued that the only valid representation of creep is one that incorporates the memory of past events. The equation of state theory, which is adopted here on the practical grounds that have been listed, does not have this feature. Unfortunately, the memory approach is not yet as well developed as is the equation of state approach. Also, the required mechanical measurements are not available in sufficient quantity to permit its adoption. All of the results reported in this book were obtained with the equation of state approach. Similar results, obtained with the memory theory approach, are simply not available at this time. The reader is advised to be aware of this and to be alert to developments concerning the memory approach. Its current status has been well described by E. Krempl [1].

To proceed with the equation of state approach we assume at the outset an expression of the form

$$\varepsilon^C = A\sigma^m t^n \tag{2.3}$$

This is the so-called Bailey-Norton law. A, m, and n are constants that are a function of temperature. The value of m is greater than one; n is usually a fraction. This law is intended to model only primary and secondary creep. Rupture is handled separately, as shown in Chapter 5. Note that the Bailey-Norton law is only one possibility. Others that are in use are given later.

For a variable stress problem, the strain rate is of interest. Hence we differentiate Eq. (2.3) with respect to time to obtain

$$\dot{\varepsilon}^C = \frac{\partial \varepsilon^C}{\partial t} = A\sigma^m n\, t^{n-1} \tag{2.4}$$

This is the so-called time hardening formulation. It states that the appropriate creep strain rate in a variable stress situation depends on stress, time and, through the constants, temperature. Note that we have ignored the time derivative of the stress in deriving the preceding formula even though we are looking for a variable stress formulation. This emphasizes one of the shortcomings of the equation of state approach and, theoretically, limits it to step changes of stress that are of long duration. In spite of this, the approach has been widely used with success and is, therefore, adopted here. Another formulation can be obtained by eliminating time from Eq. (2.4). This brings in the strain and leads to the so-called strain hardening formulation in which the creep strain rate in a variable stress situation depends on the stress, strain, and temperature. In this method Eq. (2.3) is solved for time to give

$$t = \left(\frac{\varepsilon^C}{A\sigma^m} \right)^{1/n} \tag{2.5}$$

This is now substituted into Eq. (2.4) to give

$$\dot{\varepsilon}^C = A^{1/n} n\, \sigma^{m/n} (\varepsilon^C)^{(n-1)/n} \tag{2.6}$$

Aside from the procedural differences between time hardening and strain hardening that are apparent from Eq. (2.4) and Eq. (2.6), the significance of the two formulations is demonstrated in Figs. 2.1 and 2.2. There the responses predicted by the two approaches for a given variable stress history are shown along with the creep curves for each of four constant

Uniaxial Creep Models   21

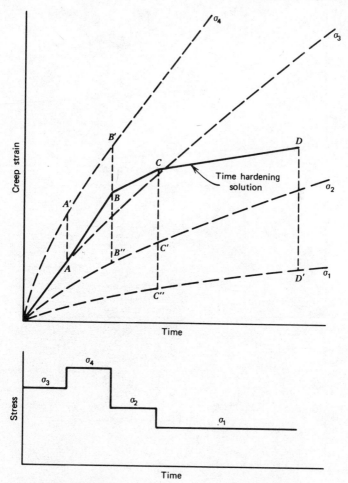

Fig. 2.1 Strain history prediction from time hardening theory. Courtesy of McGraw-Hill Company, Ltd., Design for Creep, R. K. Penny and D. L. Marriott (1971).

stress levels. In Fig. 2.1 we have the response according to the time hardening rule. At first the strain follows the σ_3 curve. When the stress shifts to σ_4 we pick up the σ_4 curve at the time of change and slide it vertically to the end of the σ_3 response. Then when the stress shifts to σ_2 we pick up the σ_2 curve at the time of the shift and slide it vertically to the end of the σ_4 response, and so on.

Now consider Fig. 2.2 where the response according to the strain hardening rule is shown. Again for σ_3 the curve starts at $t=0$. Now when the stress shifts to σ_4 we pick up the σ_4 curve at the strain value that has

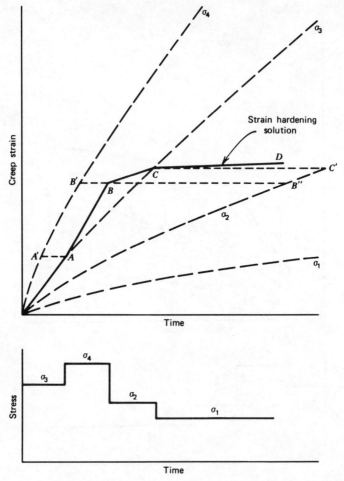

Fig. 2.2 Strain history prediction from strain hardening theory. Courtesy of McGraw-Hill Company, Ltd., Design for Creep, R. K. Penny and D. L. Marriott (1971).

been reached and slide it horizontally to the end of the σ_3 response. Then when the stress shifts to σ_2 we pick up the σ_2 curve at the strain value that has been reached and slide it horizontally to the end of the σ_4 response, and so on. Upon comparison of Fig. 2.1 and Fig. 2.2 it is evident that the two response curves are not the same. The question of which is the right one now arises. This is resolved by comparing predictions to measurements taken for a variable stress test. The results of such tests indicate that the strain hardening formulation is to be favored over the time hardening formulation. It could, however, be asked why, since both formulations stem from the same original expression, Eq. (2.4), there is any difference in results predicted? The answer to this question it that the difference is procedural, not phenomenological. If in using the time hardening rule the origin of time is adjusted appropriately at every stress increment, the two rules must by necessity give the same results. Such a procedure does not offer any advantage over the strain hardening rule. In this book we therefore adopt strain hardening. In making this choice we should also mention that two approaches have been used with strain hardening. One defines the hardening on the basis of the total strain. This is the method described in Fig. 2.2. The other defines the hardening on the basis of the primary creep strain alone. This relies on a formulation, such as the one given below, that is written in terms of primary and secondary components. The results obtained by these two approaches in predicting available experimental data do not differ much [2]. However, this may not be the case for all materials and the user is advised to be aware of the possibility of choice between the two types of strain hardening.

As an additional point it is well to emphasize that, whereas strain hardening has been adopted on the basis of its being more accurate, time hardening has not been discarded. Many argue that since there is so much scatter in creep data one might as well use the simplest possible formulation in the analysis. It is true, as we show in Section 2.6 and in Chapter 3, that the time hardening formulation is easier to solve than is the strain hardening formulation. As a result of this, time hardening forms the basis of the approximate analyses in Chapter 4 and is in fact preferred by those who seek analytical solutions. Those who pursue creep solutions on the basis of numerical analysis use the strain hardening formulation because of its demonstrated accuracy and because it is no more difficult to program for the computer than is time hardening.

As we stated before, the particular form of Eq. (2.3) that we have selected is only one of several possibilities. It is now appropriate to list some of the others that have found use, for example [2],

$$\varepsilon^C = A(\sigma)(1 - e^{-r(\sigma)t}) + \dot{\varepsilon}_m t \tag{2.7}$$

$$\varepsilon^C = \frac{a_1 t}{1 + b_1 t} + \dot{\varepsilon}_m t \tag{2.8}$$

$$\varepsilon^C = A(\sigma)(1 - e^{-r(\sigma)t}) + B(\sigma)(1 - e^{-s(\sigma)t}) + \dot{\varepsilon}_m t \tag{2.9}$$

$$\varepsilon^C = \frac{a_1 t}{1 + b_1 t} + \frac{a_2 t}{1 + b_2 t} + \dot{\varepsilon}_m t \tag{2.10}$$

These equations all separate the creep into primary and secondary creep contributions whereas the Bailey-Norton rule does not. They demonstrate that as time goes on the secondary term $\dot{\varepsilon}_m t$ predominates. There is, however, a computational difficulty with some of these since they cannot be solved explicitly for the time to permit a strain hardening formulation to be written. The time must then be obtained numerically and this is discussed in Chapter 8.

Both the time hardening and strain hardening rules that have been discussed here are applicable as long as stress reversals do not occur. If a stress reversal occurs, then auxiliary rules must be employed along with the hardening rule. An example of such an auxiliary rule is specified and explained next.

To understand the shortcomings in the strain hardening procedure when stress reversals are encountered, consider the simple case of a uniaxial creep specimen subjected first to a tensile stress of $+\sigma$ and then to a compressive stress of $-\sigma$. During the tensile portion of the loading strain hardening occurs, and the strain hardening procedure would predict that upon changing to the compressive loading, this accumulated strain hardening would be retained. For example, if the secondary creep portion of the creep response had been reached in tension, then the strain hardening procedure would predict a compressive creep strain response beginning in the secondary creep region. This appears to be incorrect. We would expect that the hardening accumulated in tension would be lost upon changing to compression, that is, the compressive creep response would exhibit primary creep similar to the case of a virgin specimen. One way of handling the problem is shown in Fig. 2.3 for a three stress history that includes a stress reversal. In Fig. 2.3(b) the creep curves for each stress level are shown. This figure assumes that the creep curve in compression is the mirror image of the creep curve in tension. This is not always the case and one should be careful to verify it for the material under consideration. In Fig. 2.3(c) the response is shown. It is determined as follows: during the period when σ_3 acts a total creep strain denoted by ε_1 is accumulated. On the basis of total creep strain, ε_1 is also a measure of the hardening. When the stress changes to $-\sigma_1$ at time t_1 it is assumed that all the hardening accumulated in tension is lost. Thus the creep response produced by the first application of compressive stress starts at zero strain hardening as in the case of a virgin specimen. Thus the $-\sigma_1$ curve used is the virgin curve. Total loss of hardening is, however, not the case for all subsequent reversals. Thus when the stress in Fig. 2.3(a) is reversed back to tension at time t_2, hardening equal to $\varepsilon_1 - \varepsilon_2$ remains in tension if ε_2, the creep strain accumulated in compression, is smaller in magnitude than ε_1. The tensile creep response then starts at this value, as shown in Fig. 2.3(c). If ε_2 is larger than ε_1 all tensile hardening is eliminated and the creep response on

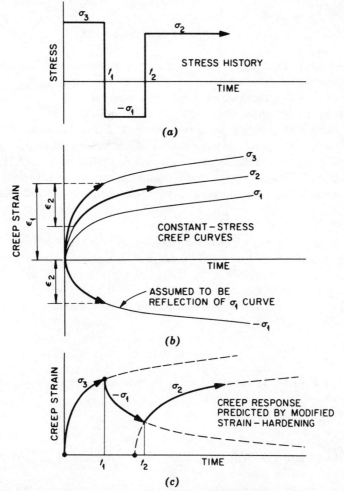

Fig. 2.3 Presumed uniaxial creep response when stress reversals are involved. Note that the strain hardening shown is based on total creep strain. Courtesy of Oak Ridge National Laboratory.

the σ_2 curve starts at zero strain. Further, if creep strain with magnitude ε_3 is accumulated at stress σ_2, and if ε_3 exceeds ε_2 in magnitude, the hardening accumulated in compression is lost. Thus a subsequent change to a compressive stress produces a creep response starting at zero strain hardening on the appropriate curve, and so on.

A general uniaxial procedure to follow is now needed. To this end, consider a general statement of the uniaxial strain hardening creep model of the form:

$$\dot{\varepsilon}^C = f(\varepsilon^H, \sigma, T) \qquad (2.11)$$

where $\dot{\varepsilon}^C$ is the total creep strain (primary plus secondary) rate and ε^H is the current strain hardening value, which in the usual strain hardening procedure is also a measure of the current creep strain.

Without modification, the uniaxial strain hardening law given by Eq. (2.11) is applicable only when stress reversals are not considered, that is, when the stresses change in magnitude but not in sign. For cyclic uniaxial loadings involving stress reversals, the applicability of Eq. (2.11) is extended by redefining the strain hardening measure, ε^H, relative to reference creep strains and by determining its value according to the following rules [2]:

1. At any time there exist two possible creep strain "origins," ε^+ and ε^-, as shown in Fig. 2.4. The strain ε^+ is a negative quantity, and ε^- is a positive quantity.
2. Initially, for a virgin specimen $\varepsilon^+ = \varepsilon^- = 0$.
3. For positive stresses, the creep rate is determined from Eq. (2.11) with ε^H defined by

$$\varepsilon^H = \varepsilon - \varepsilon^+. \qquad (2.12)$$

For negative stresses, the creep rate is determined with ε^H defined by

$$\varepsilon^H = \varepsilon - \varepsilon^-. \qquad (2.13)$$

Here ε is the current creep strain, either primary or total, depending on the strain hardening law.

4. For arbitrary stress reversals, let $\varepsilon_1, \varepsilon_2, \varepsilon_3, \ldots, \varepsilon_n$ denote the values of ε at the time of the first, second, and so on, stress reversals, respectively, and let $\varepsilon_0 = 0$ denote the initial creep strain. Then after the nth stress reversal

$$\varepsilon^+ = \min_{i=0,n} \varepsilon_i \qquad (2.14a)$$

$$\varepsilon^- = \max_{i=0,n} \varepsilon_i \qquad (2.14b)$$

The creep response predicted by these rules for a case involving ten arbitrary stress reversals is depicted in Fig. 2.4. This response is based on a

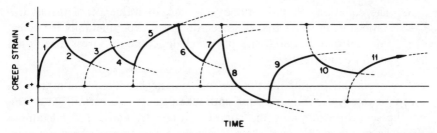

Fig. 2.4 Presumed creep response for arbitrary loading sequences. Courtesy of Oak Ridge National Laboratory.

constant stress acting between each reversal. The circular points indicate the creep strain origin used for the following step of the curve. In step 1, $\varepsilon = 0$ is used as the origin. In step 2, which involves a stress reversal, the origin ε^- is reset from its initial zero value. The creep behavior for step 2 is then obtained simply by considering the virgin uniaxial constant stress creep curve to be shifted to the new origin represented by the circular point and by reversing the direction of creep to account for the stress reversal.

The shifted origins for subsequent steps are shown by the circular points. At each stress reversal, the origin is switched so that the strain always moves away for the current origin. The origin strain, ε^+ or ε^-, is reset only when it is exceeded. In Fig. 2.4 this occurs at the beginning of steps 2, 6, and 9, and these points are called major reversal points. The remaining steps, 3, 4, 5, 7, 8, 10, and 11, begin with a residual strain hardening value determined appropriately from $(\varepsilon - \varepsilon^+)$ or $(\varepsilon - \varepsilon^-)$. Since neither strain origin, ε^+ nor ε^-, is reset at the beginning points of steps 3, 4, 5, 7, 8, 10, and 11, these points are referred to as intermediate reversal points.

2.3 MULTIAXIAL CREEP MODELS

We now turn to the derivation of a mathematical model for the creep of a solid under multiaxial stress conditions. The resulting theory must satisfy several requirements in its modeling of the creep phenomenon:

1. The multiaxial formulation must reduce to the correct uniaxial formulation when it is appropriate.
2. The model should express the constancy of volume that has been observed experimentally during the creep process.

3. The equations should embody the lack of influence of the hydrostatic state of stress that has been observed experimentally for creep.
4. For isotropic material the principal directions of stress and strain should coincide.

As we begin the development of the mathematical model we note that the creep process is dependent on the path taken to get to a certain state. This can be seen by examining the curves of Figs. 2.1 and 2.2, for example. Whereas in the elastic case there is always a one-to-one correspondence between stress and strain, in the case of creep this is not so. We can arrive at a given strain value by a single stress held constant or by a suitable variable stress program. Thus some means must be incorporated into the theory to distinguish between these two possibilities. This is accomplished by writing the theory in rate form and, thereby, providing a means for incorporating the path into the formulation. Thus it is assumed that the tensor of the creep strain rates can be written in terms of the stress deviator tensor, as follows, in a three-dimensional Cartesian coordinate system $x_i = x_1, x_2, x_3$

$$\dot{\varepsilon}_{ij}^C = \lambda S_{ij} \qquad i, j = 1, 2, 3 \tag{2.15}$$

This represents six equations. λ is a factor of proportionality that is determined below. The stress deviator tensor S_{ij} is determined from the stress tensor σ_{ij} by the definition

$$S_{ij} = \sigma_{ij} - \frac{1}{3}\sigma_{kk}\delta_{ij} \tag{2.16}$$

where δ_{ij} is the Kronecker delta. It equals zero when $i \neq j$ and one when $i = j$. The quantity $\sigma_{kk}/3$ is the sum of the three normal stresses divided by three. It is also referred to as the hydrostatic stress. This is because it is equivalent to one value of a normal stress that acts in all directions at a point, as would be the case with hydrostatic pressure. The stress deviator represents the distortional component of the stress since it is formed by subtracting the hydrostatic or volumetric portion of the stress from the stress tensor. If we write the stress tensor as a matrix

$$\sigma_{ij} = \begin{bmatrix} \sigma_{11} & \sigma_{12} & \sigma_{13} \\ \sigma_{12} & \sigma_{22} & \sigma_{23} \\ \sigma_{13} & \sigma_{23} & \sigma_{33} \end{bmatrix} \tag{2.17}$$

then the strains or their increments can be similarly expressed. Moreover, the components of the stress deviator tensor take the form

$$S_{ij} = \begin{bmatrix} \frac{1}{3}(2\sigma_{11}-\sigma_{33}-\sigma_{22}) & \sigma_{12} & \sigma_{13} \\ \sigma_{12} & \frac{1}{3}(2\sigma_{22}-\sigma_{33}-\sigma_{11}) & \sigma_{23} \\ \sigma_{13} & \sigma_{23} & \frac{1}{3}(2\sigma_{33}-\sigma_{11}-\sigma_{22}) \end{bmatrix}$$

(2.18)

The use of the deviator satisfies the third basic requirement that the formulation be independent of hydrostatic pressure. The stress-strain law [Eq. (2.15)] satisfies the fourth requirement that the stress and strain components be colinear. Equation (2.15) is known as the flow rule. It is an extension to creep of a similar rule for plasticity that was proposed by Prandtl and Reuss [3]. It is also similar to Hooke's law for elastic solids when it is expressed in deviator form

$$e_{ij}^E = \frac{1}{2\mu} S_{ij} \qquad (2.19)$$

where μ is the shear modulus and e_{ij}^E is the deviator of the elastic strain components, or

$$e_{ij} = \varepsilon_{ij} - \frac{1}{3}\varepsilon_{kk}\delta_{ij} \qquad (2.20)$$

Note that, as stated above, in the elastic case increments are not used since there is always one stress value for each strain value. Now let us determine λ in Eq. (2.15). First we define an effective stress

$$\sigma_e = \left(\frac{1}{\sqrt{2}}\right)\left[(\sigma_{11}-\sigma_{22})^2 + (\sigma_{33}-\sigma_{22})^2 + (\sigma_{11}-\sigma_{33})^2 + 6(\sigma_{12}^2+\sigma_{23}^2+\sigma_{13}^2)\right]^{1/2}$$

(2.21)

This is also expressible in the alternate form

$$\sigma_e = \sqrt{3J_2}, \qquad J_2 = \frac{S_{ij}S_{ij}}{2} \qquad (2.22)$$

Here J_2 is the second invariant of the stress deviator tensor and is defined as shown.

We also define an effective creep strain rate

$$\dot{\bar{\varepsilon}}^C = \left(\frac{\sqrt{2}}{3}\right)\{(\dot{\varepsilon}_{11}^C - \dot{\varepsilon}_{22}^C)^2 + (\dot{\varepsilon}_{22}^C - \dot{\varepsilon}_{33}^C)^2 + (\dot{\varepsilon}_{33}^C - \dot{\varepsilon}_{11}^C)^2$$
$$+ 6[(\dot{\varepsilon}_{12}^C)^2 + (\dot{\varepsilon}_{13}^C)^2 + (\dot{\varepsilon}_{23}^C)^2]\}^{1/2} \tag{2.23}$$

which is also expressible as

$$\dot{\bar{\varepsilon}}^C = \sqrt{\frac{4I_2}{3}}, \quad I_2 = \frac{\dot{\varepsilon}_{ij}^C \dot{\varepsilon}_{ij}^C}{2} \tag{2.24}$$

where I_2 is the second invariant of the strain rate tensor, and is defined as above. Now if we substitute the creep law Eq. (2.15) into Eq. (2.23) for the effective creep strain rate and observe Eq. (2.21) for the effective stress we find that

$$\lambda = \frac{3}{2\sigma_e} \frac{d\bar{\varepsilon}^C}{dt} \tag{2.25}$$

This quantity is obtained experimentally from a uniaxial creep test. We now observe that the use of the effective quantities is intended to satisfy the first requirement that the multiaxial formulation must reduce to the correct uniaxial formulation. If we consider a uniaxial case with $\sigma_{11} \neq 0$ and the other stress components all equal to zero, then Eq. (2.21) reduces to $\sigma_e = \sigma_{11}$. Furthermore, in the uniaxial case $\dot{\varepsilon}_{11}^C \neq 0$, $\dot{\varepsilon}_{22}^C = \dot{\varepsilon}_{33}^C$, and the rest of the creep rates are zero. Then by constancy of volume in creep we have

$$\dot{\varepsilon}_{11}^C + \dot{\varepsilon}_{22}^C + \dot{\varepsilon}_{33}^C = 0 \tag{2.26}$$

from which

$$\dot{\varepsilon}_{22}^C = \dot{\varepsilon}_{33}^C = -\tfrac{1}{2}\dot{\varepsilon}_{11}^C \tag{2.27}$$

Now, if we substitute Eq. (2.27) into Eq. (2.23) we find that it reduces to $\dot{\bar{\varepsilon}}^C = \dot{\varepsilon}_{11}^C$.

Finally, we may check to see if Eqs. (2.15) satisfy the second requirement on constancy of volume during creep. To do this we add the normal components from Eq. (2.15),

$$\dot{\varepsilon}_{11}^C + \dot{\varepsilon}_{22}^C + \dot{\varepsilon}_{33}^C = \lambda(S_{11} + S_{22} + S_{33}) = 0$$

as will be seen if the normal components of Eq. (2.18) are summed. Thus all of the basic requirements have been satisfied through the choice of Eq. (2.15) and the definitions of the effective stress and strain by Eqs. (2.21) and (2.23).

Now let us return to Eq. (2.25) for λ. This is obtained from a uniaxial creep test and a hardening rule for variable stress. We have discussed these matters in Section 2.2. Therefore we recall that

$$\varepsilon^C = A\sigma^m t^n$$

from Eq. (2.3). This is now extended to the multiaxial case by introducing the effective quantities, or

$$\bar{\varepsilon}^C = A\sigma_e^m t^n \tag{2.28}$$

since the effective quantities have been defined so as to reduce to the proper uniaxial result when appropriate. So now we may write

$$\lambda = \frac{3}{2\sigma_e} \frac{d\bar{\varepsilon}^C}{dt} = \frac{3}{2} A n \sigma_e^{m-1} t^{n-1} \tag{2.29}$$

and use whatever hardening rule we desire. Thus time hardening is represented by

$$\dot{\varepsilon}_{ij}^C = \tfrac{3}{2} S_{ij} A n \sigma_e^{m-1} t^{n-1} \tag{2.30}$$

and strain hardening is represented by

$$\dot{\varepsilon}_{ij}^C = \tfrac{3}{2} S_{ij} A^{1/n} n \sigma_e^{(m/n)-1} (\bar{\varepsilon}^C)^{(n-1)/n} \tag{2.31}$$

This completes the derivation of the multiaxial flow rules. The form of Eq. (2.31) is only one of the possibilities available. In some cases it is rewritten with an equivalent set of parameters defined as follows:

$$-\mu \equiv \frac{n-1}{n}, \qquad \frac{m}{n} \equiv m$$

$$nA^{1/n} \equiv \frac{1}{\tau \sigma_m^m} \tag{2.32}$$

With these exchanges Eq. (2.31) takes the alternate form

$$\dot{\varepsilon}_{ij}^C = \frac{1}{\tau} \left(\frac{\sigma_e}{\sigma_m}\right)^m (\bar{\varepsilon}^C)^{-\mu} \frac{3}{2} \frac{S_{ij}}{\sigma_e} \tag{2.33}$$

The effective creep strain increment may now be related to the effective stress as follows:

$$\bar{\dot{\varepsilon}}^C = \left(\tfrac{2}{3}\dot{\varepsilon}^C_{ij}\dot{\varepsilon}^C_{ij}\right)^{1/2}$$

$$\dot{\varepsilon}^C_{ij} = \tfrac{3}{2}F\frac{S_{ij}}{\sigma_e}, \qquad F \equiv \frac{1}{\tau}\left(\frac{\sigma_e}{\sigma_m}\right)^m (\bar{\varepsilon}^C)^{-\mu} \qquad (2.34)$$

Thus by substitution we find that

$$\bar{\dot{\varepsilon}}^C = F \qquad (2.35)$$

or

$$\frac{d\bar{\varepsilon}^C}{dt} = \frac{1}{\tau}\left(\frac{\sigma_e}{\sigma_m}\right)^m (\bar{\varepsilon}^C)^{-\mu}$$

By algebraic manipulation we can also show that this leads to

$$\frac{d}{dt}\left((\bar{\varepsilon}^C)^{1+\mu}\right) = \frac{1+\mu}{\tau}\left(\frac{\sigma_e}{\sigma_m}\right)^m$$

which upon integration gives

$$\bar{\varepsilon}^C = \left\{\frac{1+\mu}{\tau}\int_0^t \left(\frac{\sigma_e}{\sigma_m}\right)^m d\eta\right\}^{1/1+\mu} \qquad (2.36)$$

If this is applied to a case of constant uniaxial stress σ, it reduces to

$$\varepsilon^C = \left(\frac{1+\mu}{\tau}\left(\frac{\sigma}{\sigma_m}\right)^m t\right)^{1/1+\mu} \qquad (2.37)$$

and this can be used to fit the data from a uniaxial creep test. Note that τ is arbitrary since it could be merged with σ_m^m without loss of generality. Thus Eq. (2.37) is used to find m, μ, and $\tau\sigma_m^m$.

2.4 FORMULATION OF THE CREEP PROBLEM

At this point it is useful to bring together all of the equations that would be required in the formulation of a problem in creep analysis. In writing these equations we assume that in the general case the total strain is made up of elastic, plastic, creep, and thermal components. We assume that the reader is familiar with plasticity. If not, various references on plasticity that are given at the end of the chapter should be consulted, or else the plasticity contributions should be ignored. They are only included for completeness.

Thus we note first that the total strain can be decomposed as follows:

$$\varepsilon_{ij} = \varepsilon_{ij}^E + \varepsilon_{ij}^P + \varepsilon_{ij}^C + \alpha T \delta_{ij} \tag{2.38}$$

where the superscripts E, P, and C refer to the elastic strain, plastic strain, and creep strain, respectively. α is the coefficient of linear thermal expansion, T is the temperature change, and δ_{ij} is the Kronecker delta. The latter is included to indicate that only normal strains are induced by temperature changes, that is, the solid is isotropic. Next, we write Hooke's law, which always governs the elastic components of strain:

$$e_{ij}^E = \frac{1}{2\mu} S_{ij} \tag{2.39}$$

This is Eq. (2.19) repeated. Now we write the creep flow rules that have just been derived, that is,

$$\dot{\varepsilon}_{ij}^C = \frac{3}{2} \frac{d\bar{\varepsilon}^C}{dt} \frac{S_{ij}}{\sigma_e} \tag{2.40}$$

and the experimentally determined creep curve

$$\bar{\varepsilon}^C = \bar{\varepsilon}^C(\sigma_e, T, t) \tag{2.40a}$$

where

$$\sigma_e = \sqrt{(\tfrac{3}{2}) S_{ij} S_{ij}} \quad \text{and} \quad d\bar{\varepsilon}^C = \sqrt{(\tfrac{2}{3}) d\varepsilon_{ij}^C d\varepsilon_{ij}^C} \tag{2.41}$$

The effective creep strain is similarly defined.

The plastic flow rules are given by [3]

$$d\varepsilon_{ij}^P = \frac{3}{2} \frac{d\bar{\varepsilon}^P}{\sigma_e} S_{ij} \tag{2.42}$$

and the experimentally determined instantaneous stress-strain curve

$$\bar{\varepsilon}^P = \bar{\varepsilon}^P(\sigma_e) \tag{2.42a}$$

where

$$d\bar{\varepsilon}^P = \sqrt{\left(\tfrac{2}{3}\right) d\varepsilon_{ij}^P d\varepsilon_{ij}^P} \qquad (2.43)$$

is the effective plastic strain increment. The effective plastic strain $\bar{\varepsilon}^P$ is similarly defined. Equation (2.42) is valid only if the body is being loaded and has yielded, for example, if

$$\sigma_e \geqslant \sigma_y \qquad (2.44)$$

where σ_y is the yield stress in tension. This is the so-called von Mises yield condition for multiaxial stress states. If the body is unloading or if $\sigma_e < \sigma_y$, then no plastic strains occur and $d\varepsilon_{ij}^P = 0$. Further details of the mathematical theory of plasticity are not given. An excellent description can be found in Mendelson's text [3].

Now we add the equations that govern a solid no matter what type of behavior it undergoes ($i, j = 1, 2, 3$):

Equilibrium:

$$\frac{\partial \sigma_{ij}}{\partial x_i} + f_j = 0 \qquad (2.45)$$

where f_j is a body force vector.

Strain Displacement Equation:

$$\varepsilon_{ij} = \frac{1}{2}\left(\frac{\partial u_i}{\partial x_j} + \frac{\partial u_j}{\partial x_i}\right) \qquad (2.46)$$

where small strains have been assumed but are not required and u_i are the displacements in the three coordinate directions.

Heat Conduction Equation:

$$\nabla^2 T + \frac{q'''}{k} = \frac{1}{\kappa}\frac{\partial T}{\partial t} \qquad (2.47)$$

where q''', k, and κ are a heat source term, if any, thermal conductivity, and thermal diffusivity, respectively. ∇^2 is Laplace's operator

$$\nabla^2(\ ) = \frac{\partial^2(\ )}{\partial x_1^2} + \frac{\partial^2(\ )}{\partial x_2^2} + \frac{\partial^2(\ )}{\partial x_3^2}$$

Stress and Strain Deviator Definitions:

$$S_{ij} = \sigma_{ij} - \tfrac{1}{3}\sigma_{kk}\delta_{ij} \qquad (2.48a)$$

$$e_{ij} = \varepsilon_{ij} - \tfrac{1}{3}\varepsilon_{kk}\delta_{ij} \qquad (2.48b)$$

In general, therefore, we have six of Eqs. (2.38), six of Eqs. (2.39), seven of Eqs. (2.40), two of Eqs. (2.41), seven of Eqs. (2.42), one of Eq. (2.43), three of Eqs. (2.45), six of Eqs. (2.46), one of Eq. (2.47), and twelve of Eqs. (2.48a,b), for a total of 51 equations. These govern six ε_{ij}, six ε_{ij}^E, six ε_{ij}^P, six ε_{ij}^C, six e_{ij}, six σ_{ij}, six S_{ij}, three u_i, one T, one $d\bar{\varepsilon}^P$, one $d\bar{\varepsilon}^C$, one $\bar{\varepsilon}^P$, one $\bar{\varepsilon}^C$, and one σ_e for a total of 51 unknowns. The elastic, plastic creep problem is, therefore, well set once the hardening rule is selected; the stress-strain diagram and the creep test are introduced into the plastic and creep flow rules, and boundary conditions are stated. If one does not wish to consider plasticity, seven of Eqs. (2.42) and one of Eq. (2.43) are deleted, as are six $d\varepsilon_{ij}^P$, one $d\varepsilon^P$, and one $\bar{\varepsilon}^P$. Thus it becomes a system of 43 equations in 43 unknowns. The formulation must also be given the shear modulus, thermal conductivity, thermal diffusivity, body forces, and heat source. As written, the formulation of the general problem indicates that while the strains depend on the temperature, the temperature does not depend on the strains. Thus the temperature can be determined independently from Eq. (2.47) and the thermal boundary and initial conditions. If the problem is isothermal, then it is not necessary to consider Eq. (2.47) and the temperature. We should also add here the rules for handling stress reversals. We gave these for the uniaxial case in Section 2.2. In particular, rules are needed to define the idea of a stress reversal when there are as many as six stress components involved. The procedure is fairly detailed and in order not to detract from the main development of the chapter it is presented in the Appendix at the end of this chapter.

2.5 RELAXATION

Although it might appear that a different formulation would be required to model relaxation of stress at constant strain than to model creep at constant stress, which has been considered so far, this does not turn out to be the case. Since fewer relaxation data are available than creep data it has been assumed by analysts that the same flow rules apply to both phenomena. That this is valid for at least one high temperature material is demonstrated in Fig. 2.5 [2]. Thus in this book we do not distinguish between flow rules for stress and relaxation. However, when applying this assumption it is advised that the reader attempt to find out if it is valid for

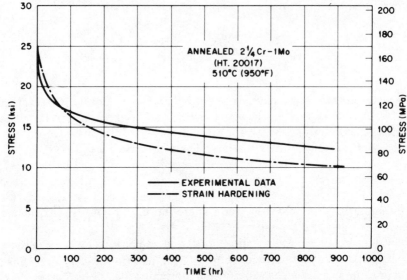

Fig. 2.5 Comparison of experimental data and strain-hardening (total creep strain) predictions for constant total strain relaxation test. Courtesy of Oak Ridge National Laboratory.

the material that is being contemplated. If such information is not available, our assumption makes a good starting point.

Similarly, we will assume that recovery is also governed by the creep flow rules that have been given here.

2.6 STEADY CREEP APPROXIMATION—ELASTIC ANALOGY

A useful approximation to the creep problem is obtained by assuming that the behavior is governed by the portion of the creep curve during which the creep rate is constant. For such situations an approximation to the creep curve that is shown in Fig. 2.6 is convenient. To represent this curve we write, for the uniaxial case,

$$\varepsilon^{CS}(\sigma, T, t) = \varepsilon_0(\sigma, T) + t\dot{\varepsilon}_m(\sigma, T) \qquad (2.49)$$

Here ε_0 is a fictitious, instantaneous strain response that is obtained by extending the constant rate portion of the creep curve back to $t=0$. It defines a fictitious elastic modulus, $E_0 = \sigma/\varepsilon_0$. $\dot{\varepsilon}_m t$ is the steady creep

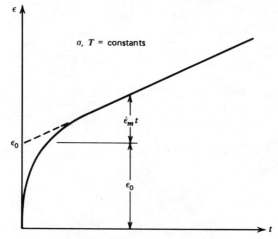

Fig. 2.6 Steady creep approximation.

strain. The steady creep strain rate is

$$\frac{d\varepsilon^{CS}}{dt} = \dot{\varepsilon}_m(\sigma, T) \tag{2.50}$$

By methods similar to those used in Section 2.3 the equation for the multiaxial case can be written

$$\frac{d\varepsilon_{ij}^{CS}}{dt} = \frac{3}{2} \frac{S_{ij}}{\sigma_e} \dot{\varepsilon}_m(\sigma_e, T) \tag{2.51}$$

Thus the creep strain rate depends only on the stress and temperature, not the time. It is said to be steady. We will use this approach to solve some problems in Chapter 3. Usually a power law is selected for the stress dependence. For the uniaxial case this gives

$$\dot{\varepsilon}_m(\sigma, T) = a(T)\sigma^m \tag{2.52}$$

with which

$$\frac{d\varepsilon^{CS}}{dt} = a(T)\sigma^m \tag{2.53}$$

Some alternate forms that have found use are

$$\frac{\dot{\varepsilon}^{CS}}{\dot{\varepsilon}_0} = \left(\frac{\sigma}{\sigma_0}\right)^m \tag{2.54}$$

and

$$\dot{\varepsilon}^{CS} = \frac{1}{\tau}\left(\frac{\sigma}{\sigma_0}\right)^m \tag{2.55}$$

In these forms the quantities $\dot{\varepsilon}_0$ and τ are not additional constants. They are merely conveniences to make the formulation dimensionless. If Eq. (2.55) is used in Eq. (2.51), we see that it is the same as would result by setting $\mu = 0$ in Eq. (2.33).

Now if we substitute Eq. (2.52) for $\dot{\varepsilon}_m(\sigma_e, T)$ in Eq. (2.51), we get the multiaxial formulation

$$\frac{d\varepsilon_{ij}^{CS}}{dt} = \tfrac{3}{2} a(T) \sigma_e^{m-1} S_{ij} \tag{2.56}$$

If we let $m = 1$, this becomes

$$\frac{d\varepsilon_{ij}^{CS}}{dt} = \tfrac{3}{2} a(T) S_{ij} \tag{2.57}$$

The elastic strain deviators due to the same S_{ij} are given by Eq. (2.39) as

$$e_{ij}^E = \frac{1}{2\mu} S_{ij} \tag{2.58}$$

Since the sum of the normal creep strain rates is zero by the constant volume requirement Eq. (2.57) can also be written for the deviator of the creep strain as

$$\frac{de_{ij}^{CS}}{dt} = \tfrac{3}{2} a(T) S_{ij} \tag{2.59}$$

Thus if we let $2\mu = 2/3a$ and replace strain by strain rate, the elasticity and steady creep solutions become the same. It has also been shown that for nonlinear elasticity [4]

$$e_{ij}^E = \tfrac{3}{2} b \sigma_e^{m-1} S_{ij} \tag{2.60}$$

where b is a constant related to Young's modulus. If we compare this to Eq. (2.56), we see that if we have a nonlinear elastic solution for b, m, and ε_{ij}^E and we switch these to a, m, and $\dot{\varepsilon}_{ij}^C$, we have a steady creep solution. Thus the stress and strain in a structure under the steady creep approximation can be found by analyzing a corresponding problem in nonlinear elasticity. This procedure is called the elastic analogy [4]. We use this in developing approximate solutions to creep problems in Chapter 4.

Finally, we may point out one further interesting point about the steady creep solution, as follows. If we redefine the time scale as

$$t = \tau^n$$

and

$$dt = n\tau^{n-1} d\tau$$

then Eq. (2.56) becomes

$$\frac{d\varepsilon_{ij}^{CS}}{d\tau} = \frac{3}{2} a \sigma_e^{m-1} n \tau^{n-1} S_{ij}$$

This is identical in form to Eq. (2.30) for the multiaxial time hardening formulation. Thus we have shown that the steady creep solution can be used to provide a time hardening solution by redefining the time scale.

2.7 CLOSURE

In this chapter we have presented the mathematical formulation of the problem of creep and, by extension, relaxation and recovery. We have done this for both uniaxial and multiaxial situations. The problem posed is a nonlinear one by virtue of the nature of the creep flow laws. It could also be geometrically nonlinear if the strains become large. As a result we will not be able to solve the problem exactly, in closed form, and we will have to appeal to some type of approximation. Thus in later chapters we present various analytical solutions that employ the steady creep and other approximations. In doing this we present problems of creep, relaxation, rupture, buckling, and cyclic loads. We also present the numerical techniques that have been applied to the creep problem for use with the digital computer.

APPENDIX: PROCEDURES FOR STRESS REVERSALS IN MULTIAXIAL CREEP

The rules for uniaxial stress reversals in creep were given in Section 2.2. In this appendix we extend them to the multiaxial case.

Many practical high temperature structural problems involve proportional loadings, that is, situations where the stresses remain in a constant proportion to each other. The extension of the rules given here for multiaxial conditions is intended primarily for application to these types of problems, and they are based on the concepts of effective stress and strain.

The general strain hardening multiaxial creep equations, analogous to Eq. (2.11) for the uniaxial case, can, by Eqs. (2.15), be written in the form

$$\dot{\varepsilon}_{ij}^C = \lambda(\bar{\varepsilon}^H, \sigma_e, T)S_{ij} \qquad (2.61)$$

where $\dot{\varepsilon}_{ij}^C$ represents the total creep strain rate components, $\bar{\varepsilon}^H$ is the current value of strain hardening (which in the usual strain hardening procedure is a measure of the current effective creep strain) on either a total or primary creep strain basis, σ_e is the effective stress, and S_{ij} represents the deviator of the stress components.

The applicability of Eq. (2.61) is extended to multiaxial stress reversals by redefining the strain hardening $\bar{\varepsilon}^H$ relative to reference creep strain rates and by determining its value according to the following generalized rules:

1. Define

Strain hardening value $\bar{\varepsilon}^H = \begin{cases} \text{Value based on } \varepsilon_{ij}^C \text{ when strain} \\ \text{hardening is based on total creep strain} \\ \text{Value based on } \varepsilon_{ij}^t \text{ when strain hardening} \\ \text{is based on primary creep strain} \end{cases}$

Instantaneous strain value $\varepsilon_{ij}^I = \begin{cases} \varepsilon_{ij}^C \text{ when strain hardening is based on} \\ \text{total creep strain} \\ \varepsilon_{ij}^t \text{ when strain hardening is based on} \\ \text{primary creep strain} \end{cases}$

$\varepsilon_{ij}^+, \varepsilon_{ij}^- = $ Two possible strain origins which exist at any time in either total creep strain space or primary creep strain space, as appropriate

$\hat{\varepsilon} = $ An effective strain quantity which in the multiaxial case is the equivalent to the distance between origins for the unixial case

Appendix: Procedures for Stress Reversals in Multiaxial Creep

$G=$ A measure, on an effective strain basis, of the distance between an instantaneous strain state and the appropriate one of the two strain origins [see Eq. (2.24)]

$$G^+ = G(\varepsilon_{ij}^I - \varepsilon_{ij}^+) = \left[\tfrac{2}{3}(\varepsilon_{ij}^I - \varepsilon^+)(\varepsilon_{ij}^I - \varepsilon_{ij}^+)\right]^{1/2}$$

$$G^- = G(\varepsilon_{ij}^I - \varepsilon_{ij}^-) = \left[\tfrac{2}{3}(\varepsilon_{ij}^I - \varepsilon_{ij}^-)(\varepsilon_{ij}^I - \varepsilon_{ij}^-)\right]^{1/2}$$

2. Definition of stress reversal: For multiaxial conditions, a "stress reversal" is considered to occur whenever the effective creep strain (G^+ or G^-) measured from the current origin (ε_{ij}^+ or ε_{ij}^-) begins to decrease. As stated previously, since the creep strain rate, or the creep strain increment, is colinear with the deviatoric stress, which is known before the creep strain increment is calculated, the condition for a stress reversal is that the deviatoric stress be directed toward the current origin. More precisely, if the current origin is ε_{ij}^+, a "stress reversal" occurs when the product

$$(\varepsilon_{ij}^I - \varepsilon_{ij}^+)S_{ij} < 0$$

Because no volume change occurs during creep, $\varepsilon_{kk}^I = 0$, and the above inequality can be replaced by

$$(\varepsilon_{ij}^I - \varepsilon_{ij}^+)\sigma_{ij} < 0$$

Similarly, if the current origin is ε_{ij}^-, a "stress reversal" occurs when

$$(\varepsilon_{ij}^I - \varepsilon_{ij}^-)\sigma_{ij} < 0$$

Whenever a load reversal is detected at the beginning of a time increment, as described below, the origin is switched (and reset if necessary) before the incremental creep strains are calculated.

3. For the initial unloaded case

$$\varepsilon_{ij}^+ = \varepsilon_{ij}^- = \hat{\varepsilon} = 0$$

4. For the initial loading of the virgin material, the creep rate is determined from Eq. (2.61) and, because of item 3 above, $\bar{\varepsilon}^H$ is defined by

$$\bar{\varepsilon}^H = G(\varepsilon_{ij}^I)$$

Assuming the initial loading is tensile in character, at the instant of the first stress reversal ε_{ij}^- and $\hat{\varepsilon}$ are set equal to ε_{ij}^I and $G(\varepsilon_{ij}^I)$, respectively, and the origin is switched to ε_{ij}^-, so that after the reversal

$$\bar{\varepsilon}^H = G^- = G(\varepsilon_{ij}^I - \bar{\varepsilon}_{ij})$$

At the instant of the next reversal, if $G^- > \hat{\varepsilon}$, then ε_{ij}^+ and $\hat{\varepsilon}$ are set equal to ε_{ij}^I and G^-, respectively, and the origin is switched to ε_{ij}^+. After each stress reversal occurs, the origin is switched.

5. In general the following steps are taken when the current origin is ε_{ij}^+ and a stress reversal occurs.

   a. If

   $$G(\varepsilon_{ij}^I - \varepsilon_{ij}^+) > \hat{\varepsilon}$$

   leave ε_{ij}^+ unchanged and reset

   $$\varepsilon_{ij}^- = \varepsilon_{ij}^I$$

   $$\hat{\varepsilon} = G(\varepsilon_{ij}^I - \varepsilon_{ij}^+)$$

   b. If

   $$G(\varepsilon_{ij}^I - \varepsilon_{ij}^+) \leq \hat{\varepsilon}$$

   leave ε_{ij}^+, ε_{ij}^-, and $\hat{\varepsilon}$ unchanged.
   c. Test for the condition discussed in step 7 below, and if it does not apply, proceed to step d.
   d. The origin is set at ε_{ij}^-, and the effective strain hardening $\bar{\varepsilon}^H$, is defined by

   $$\bar{\varepsilon}^H = G(\varepsilon_{ij}^I - \varepsilon_{ij}^-)$$

   This relation is used to determine the effective strain hardening until the next stress reversal occurs. When the next reversal occurs, proceed to step 6.

6. The following steps are taken when the current origin is ε_{ij}^- and a stress reversal occurs.

   a. If

   $$G(\varepsilon_{ij}^I - \varepsilon_{ij}^-) > \hat{\varepsilon}$$

Appendix: Procedures for Stress Reversals in Multiaxial Creep

leave ε_{ij}^- unchanged and reset

$$\varepsilon_{ij}^+ = \varepsilon_{ij}^I$$

$$\hat{\varepsilon} = G(\varepsilon_{ij}^I - \varepsilon_{ij}^-)$$

b. If

$$G(\varepsilon_{ij}^I - \varepsilon_{ij}^-) \leqslant \hat{\varepsilon}$$

leave ε_{ij}^+, ε_{ij}^-, and $\hat{\varepsilon}$ unchanged.

c. Test for the condition discussed in step 7 below, and if it does not apply, proceed to step d.

d. The origin is set at ε_{ij}^+, and the effective strain hardening, $\bar{\varepsilon}^H$, is defined by

$$\bar{\varepsilon}^H = G(\varepsilon_{ij}^I - \varepsilon_{ij}^+)$$

This relation is used to determine the effective strain hardening until the next stress reversal occurs, at which time return to step 5.

7. In most practical cases the above rules are sufficient to ensure that the creep strain increments are always directed away from the current origin. However, in either step 5 or step 6, it is possible to have a condition where both

$$(\varepsilon_{ij}^I - \varepsilon_{ij}^+)\sigma_{ij} < 0$$

and

$$(\varepsilon_{ij}^I - \varepsilon_{ij}^-)\sigma_{ij} < 0$$

and neither origin is to be reset. Figure 2.7 illustrates such a situation in which the creep strain increments along path 3 in strain space are directed toward both origins. If the rules that have been given are used, every increment taken along path 3 must be interpreted as a stress reversal and results in a switch of origins. To avoid this problem of possible repeated oscillations between origins when moving along a single path, the most distant origin should be used in such cases. That is, ε_{ij}^+ is to be used as the origin and the effective strain hardening determined as in step 5 if

$$G(\varepsilon_{ij}^I - \varepsilon_{ij}^+) \geqslant G(\varepsilon_{ij}^I - \varepsilon_{ij}^-)$$

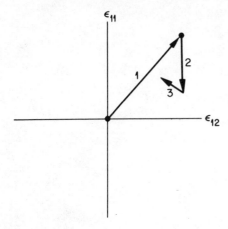

Fig. 2.7 Schematic of situations in which creep strain increments are directed toward both origins [2]. Courtesy of Oak Ridge National Laboratory.

and ε_{ij}^- is to be used as the origin and the effective strain hardening determined as in step 6 if

$$G\left(\varepsilon_{ij}^I - \varepsilon_{ij}^+\right) < G\left(\varepsilon_{ij}^I - \varepsilon_{ij}^-\right)$$

It is believed that in most practical applications the use of these rules results in reasonable and consistent predictions. It should again be pointed out, however, that we are relying on effective strain concepts, and thus we are using a single quantity—effective strain—as a repository of historical effects associated with each strain component. The shortcomings of this procedure can manifest themselves in certain proportional loading situations where anomalous strain hardening behavior can still be obtained even with the auxiliary rules. Fortunately, most practical problems involve nearly proportional loadings, and no difficulty should arise. Nonetheless, the analyst should be alert for situations that might potentially cause problems.

REFERENCES

[1] E. Krempl, Cyclic Creep—An Interpretive Literature Survey, Welding Research Council Bulletin No. 195, pp. 63–123 (1974).
[2] C. E. Pugh, Constitutive Equations for Creep Analysis of Liquid Moderated Fast Breeder Reactor (LMFBR) Components, in S. Y. Zamrik and R. I. Jetter, editors, *Advances in Design for Elevated Temperature Environment*, American Society of Mechanical Engineers, New York, (1975), pp. 1–16.
[3] A. Mendelson, *Plasticity: Theory and Application*, Macmillan Company, New York (1968).
[4] N. J. Hoff, Approximate Analysis of Structures in the Presence of Moderately Large Creep Deformation, *Q. Appl. Math.*, vol. 12, pp. 49–55 (1954).

GENERAL REFERENCES FOR FURTHER READING

Finnie, I. and W. R. Heller, *Creep of Engineering Materials*, McGraw-Hill Book Company, Inc., New York (1959).

Hult, J. A. H., *Creep in Engineering Structures*, Blaisdell Publishing Co., Waltham, Mass. (1966).

Johnson, W. and P. B. Mellor, *Engineering Plasticity*, Van Nostrand Reinhold, Co., London (1973).

Odqvist, F. K. G., *Mathematical Theory of Creep and Creep Rupture*, Clarendon Press, Oxford, England (1974).

Penny, R. K. and D. L. Marriott, *Design for Creep*, McGraw-Hill Book Co., Ltd., London (1971).

EXERCISES

2.1 Consider a bar made of a material which obeys the creep law $\varepsilon^C = A\sigma^{2.5} t^{1/3}$. Determine the final strain for time hardening and strain hardening when the uniaxial stress σ_0 acts for two hours and is followed by the stress $\sigma_0/2$ for two hours.

2.2 Consider a bar made of the material in Exercise 2.1. Now it is subjected to the uniaxial stress σ_0 for two hours, then $-\sigma_0$ for one hour and σ_0 for one hour. Use the rules for strain hardening with stress reversals to determine the final strain. Assume that the same creep law governs both tension and compression.

2.3 Derive Eq. (2.19).

2.4 How can Eq. (2.37) be used to determine μ, m, and $\tau \sigma_m^m$ from a series of creep tests?

CHAPTER 3

CREEP DEFORMATION STRESS RELAXATION AND RECOVERY PROBLEMS

3.1 INTRODUCTION

This chapter has a twofold purpose. One of its aims is to show that, in a typical creep problem with constant load, the stress state changes from an initial elastic one to an eventual, so-called stationary, creep state. The second aim is to show that the redistribution from the initial elastic state to the final stationary state, which is known as non-stationary creep, cannot in general be analyzed in closed form. This is demonstrated by considering a statically determinate problem: the thin tube, and two statically indeterminate problems: the thick tube and a two bar structure. Similar considerations are then demonstrated for relaxation and recovery problems. This involves the study of stress relaxation in a single bar and recovery in a two bar indeterminate structure. This then sets the stage for the approximate analytical techniques and the numerical approaches that are presented later in the book.

3.2 CREEP DEFORMATION PROBLEMS

3.2.1 Creep of a Thin Pressurized Tube with Closed Ends

As a first problem we consider a thin pressurized tube with closed ends. The stresses can be obtained from equilibrium and are

$$\sigma_\theta = \frac{pR_i}{h}, \qquad \sigma_z = \frac{pR_i}{2h}, \qquad \sigma_r = 0$$

Here θ, z, and r refer to the circumferential, axial, and radial directions; p is the internal pressure, and R_i and h are the inner radius and thickness of the tube. The mean stress, stress deviators, and the effective stress are found from the above to be

$$\tfrac{1}{3}\sigma_{kk} = \frac{pR_i}{2h}$$

$$S_r = -\frac{pR_i}{2h} \qquad S_\theta = \frac{pR_i}{2h} \qquad S_z = 0$$

$$\sigma_e = \frac{\sqrt{3}}{2}\frac{pR_i}{h}$$

Thus the stresses are always known. This is because the problem is statically determinate. The creep strains are obtained from the creep law. For example, if we use the strain hardening formulation in the form given by Eqs. (2.33) and (2.36), we find that for constant pressure

$$\bar{\varepsilon}^C = \left\{\frac{1+\mu}{\tau}\left(\frac{\sqrt{3}\,pR_i}{2h\sigma_m}\right)^m t\right\}^{1/(1+\mu)}$$

because σ_e is independent of time. Furthermore,

$$\dot{\varepsilon}^C_\theta = -\dot{\varepsilon}^C_r = \left\{\frac{1+\mu}{\tau}\left(\frac{\sqrt{3}\,pR_i}{2h\sigma_m}\right)^m t\right\}^{-\mu/(1+\mu)} \frac{\sqrt{3}}{2\tau}\left(\frac{\sqrt{3}\,pR_i}{2h\sigma_m}\right)^m$$

$$\dot{\varepsilon}^C_z = 0$$

Creep Deformation, Stress Relaxation, and Recovery Problems

As a result we see that the tube grows radially but maintains its length. Considering the radial displacement, for example, we have

$$\varepsilon_\theta^C = \frac{u^C}{R_i} = \int_0^t \dot\varepsilon_\theta^C \, d\eta$$

and the total radial deformation at time t is given by

$$u(t) = u_0 + R_i \int_0^t \dot\varepsilon_\theta^C \, d\eta$$

Here u_0 is the initial value and is calculated as follows (assuming elastic behavior):

$$\varepsilon_\theta^E = \frac{1}{E}(\sigma_\theta - \nu(\sigma_z + \sigma_r)) \quad \text{(Hooke's Law)}$$

$$= \frac{pR_i}{Eh}\left(1 - \frac{\nu}{2}\right)$$

Therefore,

$$u_0 = \varepsilon_\theta^E R_i = \frac{pR_i^2}{Eh}\left(1 - \frac{\nu}{2}\right)$$

and the deformation history is, after carrying out the integration of $\dot\varepsilon_\theta^C$,

$$u(t) = \frac{pR_i^2}{Eh}\left(1 - \frac{\nu}{2}\right)$$
$$+ R_i \left[\frac{1+\mu}{\tau}\left(\frac{\sqrt{3}\,pR_i}{2h\sigma_m}\right)^m\right]^{-\mu/(1+\mu)} \frac{\sqrt{3}}{2\tau}\left(\frac{\sqrt{3}\,pR_i}{2h\sigma_m}\right)^m \frac{t^{1/(1+\mu)}}{1/(1+\mu)}$$

In review of this solution we note that:

1. The stresses are constant, so the creep is called stationary.
2. The strains and the displacement are functions of time starting at $t = 0$.
3. Small strains are assumed; otherwise, changes in geometry would result in changes in stress.

Our first illustration has, therefore, been one in which an exact solution could be found. This is because the stresses were obtainable from equilibrium at all times.

3.2.2 Creep of a Thick Pressurized Tube with Closed Ends.

The pressurized thick tube is a statically indeterminate problem. Hence the stresses cannot be obtained from equilibrium alone. To show the progress of creep from the initial elastic state to the stationary state under a constant pressure, we proceed as follows:

Initial Elastic Solution. The elastic solution, known as the Lamé solution is [1]:

$$\begin{Bmatrix} \sigma_r \\ \sigma_\theta \\ \sigma_z \end{Bmatrix} = \frac{p}{(R_0/R_i)^2 - 1} \begin{Bmatrix} 1-(R_0/r)^2 \\ 1+(R_0/r)^2 \\ 1 \end{Bmatrix} \quad (3.1)$$

Here, in addition to the notation from the thin tube problem, R_0 is the outer radius of the tube and r is the radius to an arbitrary point in the wall.

Stationary Solution. We begin with the equilibrium and compatibility equations in polar coordinates [1]

$$r\frac{d\sigma_r}{dr} = \sigma_\theta - \sigma_r$$

$$r\frac{d\dot\varepsilon_\theta^C}{dr} = \dot\varepsilon_r^C - \dot\varepsilon_\theta^C$$

Here we have ignored the elastic strains in comparison to the stationary creep strains and we have written the compatibility equation in terms of rates. We now add the incompressibility of the creep rates:

$$\dot\varepsilon_\theta^C + \dot\varepsilon_r^C + \dot\varepsilon_z^C = 0$$

and the plane strain assumptions

$$\dot\varepsilon_z^C = \varepsilon_z^C = 0$$

from which, with incompressibility,

$$\dot\varepsilon_r^C = -\dot\varepsilon_\theta^C$$

Substitute this into the compatibility equation and it becomes

$$\frac{d\dot\varepsilon_r^C}{dr} = -\frac{2\dot\varepsilon_r^C}{r}$$

Integration then gives

$$\dot{\varepsilon}_r^C = -\dot{\varepsilon}_\theta^C = \frac{C}{r^2}$$

where C is an arbitrary constant. Now calculate the effective creep strain rate from Eq. (2.24). This gives

$$\dot{\bar{\varepsilon}}^C = \frac{2}{\sqrt{3}} \frac{C}{r^2} \qquad (3.1.a)$$

Now since $\dot{\varepsilon}_z^C = 0$, the flow law gives $S_z = 0$ and the effective stress is from Eq. (2.41):

$$\sigma_e = \frac{\sqrt{3}}{2}(\sigma_\theta - \sigma_r)$$

or

$$\sigma_\theta - \sigma_r = \frac{2\sigma_e}{\sqrt{3}}$$

For the stationary state the effective creep rate and effective stress are related by the uniaxial relation for steady creep

$$\frac{d\bar{\varepsilon}^C}{dt} = a\sigma_e^m$$

Therefore,

$$\sigma_\theta - \sigma_r = \frac{2}{\sqrt{3}}\left(\frac{\dot{\bar{\varepsilon}}^C}{a}\right)$$

Substitution of Eq. (3.1.a) then gives

$$\sigma_\theta - \sigma_r = \frac{2}{\sqrt{3}}\left(\frac{1}{a}\frac{2}{\sqrt{3}}\frac{C}{r^2}\right)^{1/m}$$

Substitute this into the equilibrium equation and obtain

$$\frac{d\sigma_r}{dr} = \left(\frac{2}{\sqrt{3}}\right)^{1+\frac{1}{m}}\left(\frac{C}{a}\right)^{1/m} r^{-2/m-1}$$

Integration now gives

$$\sigma_r = \frac{Dr^{-2/m}}{-2/m} + E \qquad (3.1.b)$$

where $D \equiv \left(\dfrac{2}{\sqrt{3}}\right)^{1+1/m}\left(\dfrac{C}{a}\right)^{1/m}$ and E are arbitrary constants. These are found from the boundary conditions

$$\sigma_r = -p \quad \text{at } r = R_i$$
$$\sigma_r = 0 \quad \text{at } r = R_0$$

From these conditions we find

$$D = \frac{2p}{m}\left(R_i^{-2/m} - R_0^{-2/m}\right)^{-1}$$

$$E = pR_0^{-2/m}\left(R_i^{-2/m} - R_0^{-2/m}\right)^{-1}$$

Now we note from equilibrium that

$$\sigma_\theta = \sigma_r + r\frac{d\sigma_r}{dr} \qquad (3.1.c)$$

and from $S_z = 0$

$$\sigma_z = \tfrac{1}{2}(\sigma_r + \sigma_\theta) \qquad (3.1.d)$$

Thus substituting E and D into Eq. (3.1.b) and then using Eqs. (3.1.c) and (3.1.d) gives the stationary stress solution as

$$\left\{\begin{array}{c}\sigma_r \\ \sigma_\theta \\ \sigma_z\end{array}\right\} = \frac{p}{(R_0/R_i)^{2/m}-1}\left\{\begin{array}{c}1-\left(\dfrac{R_0}{r}\right)^{2/m} \\ 1+\dfrac{2-m}{m}\left(\dfrac{R_0}{r}\right)^{2/m} \\ 1+\dfrac{1-m}{m}\left(\dfrac{R_0}{r}\right)^{2/m}\end{array}\right\} \qquad (3.2)$$

Note that these are independent of time.

A comparison of the stationary solution Eq. (3.2) with the elastic solution Eq. (3.1) shows some interesting points. First of all we note that when m is set to unity the elastic solution results. This is in accord with the

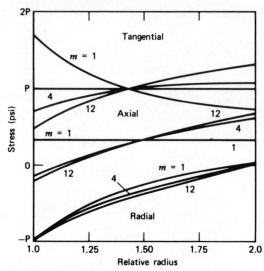

Fig. 3.1 Stress redistribution in a thick cylinder under internal pressure [2]. Courtesy of McGraw-Hill Book Co., Inc. I. Finnie and W. R. Heller, Creep of Engineering Materials, (1959).

elastic analogy that was considered in Section 2.6. Other interesting results can be found by plotting the two sets of equations for various values of m. We choose $R_0/R_i = 2$ and give the results in Fig. 3.1 [2]. There the circumferential, axial, and radial stresses are plotted against the relative radius as multiples of the pressure p for various values of m. First, we note that for $m = 1$, say, the maximum circumferential stress is on the inner surface while for any $m > 1$ the maximum circumferential stress is on the outer surface. This has serious ramifications in the evaluation of the maximum stress in the tube, as will be seen Chapter 5 on creep rupture. Second, we note that $m = 1$ is the elastic solution and $m > 1$ is the final stationary solution for the value of m that applies. Thus we see that redistribution has occurred. That is to say, a non-stationary creep solution has to be carried out to see how the initial elastic stress state redistributes into the final stationary state. This cannot be done in closed form. The non-stationary redistribution is a direct consequence of the static indeterminacy of the problem. Finally, if we wish to know the stationary creep strain distribution, we would use the flow law for steady creep as given by Eq. (2.56).

3.2.3 A Statically Indeterminate Two Bar Structure

Now let us examine the relatively simple two bar structure subjected to an axial load as shown in Fig. 3.2 [3]. We wish to determine the stress and

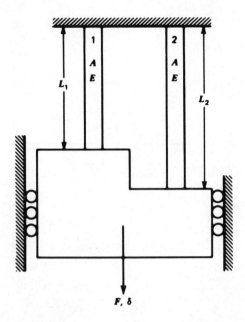

Fig. 3.2 Indeterminate two bar structure [3].

strain history. The bars are attached to a rigid block that is arranged so that the load F produces only axial forces, not bending, in each bar. The two bars are of the same material and area, and differ only in their length. We shall show that even for this relatively simple problem the non-stationary creep solution cannot be obtained in closed form. As before, we will start with the initial elastic solution, then we will find the stationary solution, and finally we shall attempt the non-stationary solution.

Initial Elastic Solution. We start with the uniaxial elastic law for each bar:

$$\varepsilon_i = \sigma_i/E \qquad i = 1, 2 \qquad (3.3)$$

where 1, and 2 refer to the two bars. By axial force equilibrium we have

$$F = A(\sigma_1(0) + \sigma_2(0)) \qquad (3.4)$$

where $\sigma_1(0)$, and $\sigma_2(0)$ are the initial stresses in bar 1 and bar 2, respectively, and A is the area of each bar. The strain displacement equations are

$$\delta(0) = \varepsilon_1(0) L_1 = \varepsilon_2(0) L_2 \qquad (3.5)$$

where $\varepsilon_1(0)$, L_1 and $\varepsilon_2(0)$, L_2 are the initial axial strain and length for each of bar 1 and bar 2. Both axial displacements are $\delta(0)$, the overall initial displacement, by virtue of the rigid yoke. Now substitute the elastic law

Eq. (3.3) into the equilibrium equation Eq. (3.4). This gives

$$\varepsilon_1(0) + \varepsilon_2(0) = F/AE \tag{3.6}$$

Then in view of Eq. (3.5)

$$\delta(0)\left(\frac{1}{L_1} + \frac{1}{L_2}\right) = \frac{F}{AE} \tag{3.7}$$

From this, Eqs. (3.3) and (3.5) give the stresses as

$$\sigma_i(0) = \varepsilon_i(0)E = \frac{\delta(0)}{L_i}E \quad i=1,2 \tag{3.8}$$

Upon substitution of $\delta(0)$ the initial elastic solution becomes

$$\delta(0) = \frac{L_1 L_2}{L_1 + L_2} \frac{F}{AE}$$

$$\sigma_1(0) = \frac{L_2}{L_1 + L_2} \frac{F}{A} \tag{3.9}$$

$$\sigma_2(0) = \frac{L_1}{L_1 + L_2} \frac{F}{A}$$

Stationary Creep Solution. Now we start with the uniaxial steady creep law:

$$\dot{\varepsilon}_i^{CS} = a\sigma_i^m \quad i=1,2 \tag{3.10}$$

From this,

$$\sigma_i = \left(\frac{\dot{\varepsilon}_i^{CS}}{a}\right)^{1/m} \quad i=1,2 \tag{3.11}$$

The equilibrium Eq. (3.4) still holds, therefore

$$\left(\frac{\dot{\varepsilon}_1^{CS}}{a}\right)^{1/m} + \left(\frac{\dot{\varepsilon}_2^{CS}}{a}\right)^{1/m} = \frac{F}{A} \tag{3.12}$$

The strain displacement Eqs. (3.5) are also valid. Hence,

$$\left(\frac{\dot{\delta}_{CS}}{a}\right)^{1/m}\left(\frac{1}{L_1^{1/m}}+\frac{1}{L_2^{1/m}}\right)=\frac{F}{A} \tag{3.13}$$

Furthermore,

$$\sigma_i^{CS}=\left(\frac{\dot{\delta}_{CS}}{L_i a}\right)^{1/m} \quad i=1,2 \tag{3.14}$$

from the strain rate-displacement rate form of Eq. (3.5) and the steady creep law Eq. (3.10). Now if we solve Eq. (3.13) for $\dot{\delta}_{CS}$ and substitute it into the foregoing, we obtain the stationary creep solution

$$\dot{\delta}_{CS}=a\left(\frac{F}{A}\right)^m\frac{(L_1 L_2)}{(L_2^{1/m}+L_1^{1/m})^m}$$

$$\sigma_1^{CS}=\frac{L_2^{1/m}}{L_1^{1/m}+L_2^{1/m}}\frac{F}{A} \tag{3.15}$$

$$\sigma_2^{CS}=\frac{L_1^{1/m}}{L_1^{1/m}+L_2^{1/m}}\frac{F}{A}$$

Now let us compare the stationary creep solution Eq. (3.15) to the initial elastic solution Eq. (3.9). First we note that if $m=1$, the two solutions coincide if the displacement rate is replaced with the displacement and a is replaced with $1/E$ (see the elastic analogy in Section 2.6). Second, we observe that the stresses in each bar change; that is, $\sigma_1(0)\neq\sigma_1^{CS}$ and $\sigma_2(0)\neq\sigma_2^{CS}$. Also, the ratio of stresses in the two bars changes from an initial value to a final value, that is

$$\text{from } \frac{\sigma_1(0)}{\sigma_2(0)}=\frac{L_2}{L_1} \text{ to } \frac{\sigma_1^{CS}}{\sigma_2^{CS}}=\left(\frac{L_2}{L_1}\right)^{1/m} \tag{3.16}$$

Thus we see that the static indeterminacy of this problem causes stress redistribution to occur.

Non-stationary Creep Solutions. Now we need to consider the total strain to be made up of both elastic and creep components according to

$$\varepsilon_i=\varepsilon_i^E+\varepsilon_i^C \quad i=1,2 \tag{3.17}$$

The strain displacement Eqs. (3.5) and the equilibrium Eq. (3.4) still hold and the elastic law Eq. (3.3) governs ε_i^E. The uniaxial creep strain rate in each bar is given by

$$\dot{\varepsilon}_i^C = B\sigma_i^m n t^{n-1} \tag{3.18}$$

assuming time hardening. B is used to avoid confusion with the area A.

Now since $\sigma_2 = E\varepsilon_2^E$, the equilibrium Eq. (3.4) gives

$$\varepsilon_2^E = \left(\frac{F}{A} - \sigma_1\right)\frac{1}{E} \tag{3.19}$$

Therefore the equality of axial displacements becomes

$$L_1\left(\varepsilon_1^C + \frac{\sigma_1}{E}\right) = L_2\left[\varepsilon_2^C + \left(\frac{F}{A} - \sigma_1\right)\frac{1}{E}\right] \tag{3.20}$$

where the quantity in the parentheses on the left is the total strain ε_1 in bar 1 and the quantity in the brackets is the total strain ε_2 in bar 2. We now differentiate Eq. (3.20) and assume that the lengths and areas remain essentially constant. This yields

$$L_1\left(\dot{\varepsilon}_1^C + \frac{\dot{\sigma}_1}{E}\right) = L_2\left(\dot{\varepsilon}_2^C - \frac{\dot{\sigma}_1}{E}\right) \tag{3.21}$$

since the force is constant. On rearrangement this becomes

$$\frac{L_1 + L_2}{E}\frac{d\sigma_1}{dt} = L_2\dot{\varepsilon}_2^C - L_1\dot{\varepsilon}_1^C \tag{3.22}$$

Finally, the creep law is substituted to give

$$\frac{L_1 + L_2}{E}\frac{d\sigma_1}{dt} = L_2 B n t^{n-1}\left(\frac{F}{A} - \sigma_1\right)^m - L_1 B \sigma_1^m n t^{n-1} \tag{3.23}$$

This cannot be solved in closed form even though it represents the solution of a rather simple problem. We have presented it to motivate our later presentation of approximate and numerical methods.

To close this first presentation of creep problems we reiterate our finding that, whenever the load is constant and the problem is statically indeterminate, there is a redistribution of stress from the initial state to a final stationary state. If the system is statically determinate, then the stresses are always known and there is no redistribution due to a steady load. The strains in such a case follow directly from the flow rule.

3.3 A STRESS RELAXATION PROBLEM

In the previous section we prescribed the load and observed the creep strains and stresses. Now let us consider the situation where the displacement is prescribed and the stresses that brought them about relax.

As an example, we consider a bar that has been elongated an amount δ by a force P. We wish to determine how P must change to maintain δ. The bar is of length L.

The total uniaxial strain is

$$\varepsilon = \frac{\delta}{L} = \varepsilon^E + \varepsilon^P + \varepsilon^C \tag{3.24}$$

where the superscripts E, P, and C denote the elastic, plastic, and creep components. Here, as will be seen shortly, it is a simple matter to include the plastic strain. At constant displacement δ the strain rate satisfies

$$\dot{\varepsilon}^E + \dot{\varepsilon}^P + \dot{\varepsilon}^C = 0 \tag{3.25}$$

Since the bar unloads during relaxation, it is safe to say that $\dot{\varepsilon}^P = 0$. This is in accordance with our knowledge, from Chapter 1, that there is no change in plastic strain during unloading. Further, if we use a steady creep law as a first approximation we obtain

$$\frac{1}{E_0} \frac{d\sigma}{dt} + a\sigma^m = 0 \tag{3.26}$$

Rearrangement gives

$$\frac{d\sigma}{\sigma^m} = -E_0 a\, dt \tag{3.27}$$

and integration gives

$$\frac{\sigma^{1-m}}{1-m} = -E_0 a t + C \tag{3.28}$$

where C is an arbitrary constant. It is found from the initial condition:

$$\text{At } t=0, \quad \sigma = \sigma_0 \quad \therefore C = \frac{\sigma_0^{1-m}}{1-m}$$

The resulting expression for the stress is

$$\sigma^{1-m} = \sigma_0^{1-m} + (m-1) E_0 a t \tag{3.29}$$

58   Creep Deformation, Stress Relaxation, and Recovery Problems

provided that $m \neq 1$. To find σ_0 we note that

$$\varepsilon = \frac{\delta}{L} = \frac{\sigma_0}{E_0}$$

from which

$$\sigma_0 = \frac{E_0 \delta}{L} \tag{3.30}$$

Recall that E_0 is found from the $t=0$ intercept of the steady creep approximation that was discussed in Section 2.6. Now if we substitute this into the stress expression and introduce $\sigma = P/A$, we obtain an equation for the force

$$P = A\left(\left(\frac{L}{E_0 \delta}\right)^{m-1} + (m-1)E_0 t a\right)^{-1/(m-1)} \tag{3.31}$$

Thus as t increases, P decreases. This result was based upon the steady creep approximation. If we use the strain hardening creep law we should have to solve

$$\frac{1}{E}\frac{d\sigma}{dt} + B^{1/n}\sigma^{m/n} n (\varepsilon^C)^{(n-1)/n} = 0 \tag{3.32}$$

instead of Eq. (3.26). Since

$$\varepsilon^C = \varepsilon - \varepsilon^E = \frac{\delta}{L} - \frac{\sigma}{E} = \frac{\sigma_0 - \sigma}{E} \tag{3.33}$$

the governing equation becomes

$$\frac{1}{E}\frac{d\sigma}{dt} + B^{1/n}\sigma^{m/n} n \left(\frac{\sigma_0 - \sigma}{E}\right)^{(n-1)/n} = 0 \tag{3.34}$$

which is more difficult to solve. This is the same state of affairs as we found in the creep case. Suppose we try the time hardening rule. Now the equation to solve becomes, with the actual value of the elastic modulus,

$$\frac{1}{E}\frac{d\sigma}{dt} + B\sigma^m n t^{n-1} = 0 \tag{3.35}$$

From this, rearrangement gives

$$\frac{d\sigma}{\sigma^m} = -EBn t^{n-1} dt \tag{3.36}$$

Integration then gives

$$\frac{\sigma^{1-m}}{1-m} = -EBt^n + C_1 \tag{3.37}$$

Here C_1 is again an arbitrary constant. It is found by application of the previous initial conditions to be

$$C_1 = \frac{\sigma_0^{1-m}}{1-m}$$

Thus the stress expression becomes

$$\sigma^{1-m} = \sigma_0^{1-m} + (m-1)EBt^n \tag{3.38}$$

Now compare this to Eq. (3.29). If in accordance with Section 2.6 we let

$$\tau = t^n \quad (\text{and } E_0 = E,\ B = a)$$

the foregoing expression becomes identical to Eq. (3.29) except for the time definition. Thus we see that a time hardening solution can be found from a steady creep solution, as discussed in Section 2.6.

3.4 RECOVERY IN A STATICALLY INDETERMINATE TWO BAR STRUCTURE

Now let us return to the two bar structure with rigid yoke that we considered as a creep problem in Section 3.2.3. This time we assume that the stresses have reached the stationary state given by Eqs. (3.15) and remove the force F. The aim is to determine the resulting state of stress in each bar. The immediate result of removing the load is that the elastic stresses, Eqs. (3.9), are recovered, that is, subtracted from the stationary stresses, Eqs. (3.15). Thus for the present problem the initial stress state, upon removing F, is

$$\sigma_1(0) = \frac{L_2^{1/m}}{L_1^{1/m} + L_2^{1/m}} \frac{F}{A} - \frac{L_2}{L_1 + L_2} \frac{F}{A}$$

$$\sigma_2(0) = \frac{L_1^{1/m}}{L_1^{1/m} + L_2^{1/m}} \frac{F}{A} - \frac{L_1}{L_1 + L_2} \frac{F}{A} \tag{3.39}$$

Note that this distribution satisfies the equilibrium condition for the present problem, that is, $\sigma_1+\sigma_2=0$. In addition, we have

$$\varepsilon_i=\delta/L_i \qquad i=1,2 \tag{3.39a}$$

where δ is now the same for each bar but varies with time. By strain decomposition

$$\varepsilon_i=\varepsilon_i^E+\varepsilon_i^C, \qquad \frac{d\varepsilon_i}{dt}=\dot{\varepsilon}_i^E+\dot{\varepsilon}_i^C \tag{3.40}$$

and using both the elastic and the steady creep laws,

$$\varepsilon_i^E=\frac{\sigma_i}{E_0}, \qquad \dot{\varepsilon}_i^C=a\sigma_i^m \tag{3.40a}$$

Thus by combining the foregoing

$$\frac{1}{L_i}\frac{d\delta}{dt}=a\sigma_i^m+\frac{1}{E_0}\frac{d\sigma_i}{dt} \qquad i=1,2 \tag{3.41}$$

This gives two equations in the three unknowns σ_1, σ_2, and δ. If we subtract one from the other and note that $\sigma_2=-\sigma_1$, we obtain

$$\frac{1}{E_0}(L_1+L_2)\frac{d\sigma_1}{dt}+\sigma_1^m(L_1-(-1)^m L_2)a=0 \tag{3.41a}$$

This integrates to give

$$\frac{\sigma_1^{1-m}}{1-m}=\frac{E_0 a(L_2(-1)^m-L_1)}{L_2+L_1}t+C_2 \tag{3.42}$$

C_2 is an arbitrary constant that is found from the initial condition on σ_1 to be

$$C_2=\frac{1}{1-m}\left(\frac{L_2^{1/m}}{L_1^{1/m}+L_2^{1/m}}\frac{F}{A}-\frac{L_2}{L_1+L_2}\frac{F}{A}\right)^{1-m} \tag{3.42a}$$

where F is the original force that produced the initial state. With this we find

$$\sigma_1=\left\{(1-m)C_2+\frac{E_0 a(L_2(-1)^m-L_1)}{L_2+L_1}t\right\}^{1/(1-m)} \tag{3.43}$$

and a similar equation for σ_2. If we wanted to obtain a time hardening solution we would replace t with t^m, E_0 with E and a with B. It is interesting to note, in conjunction with the initial stresses Eq. (3.39), that whereas both bars were originally in tension during the stationary state Eqs. (3.15), the immediate result of removing the load F is to put one bar into compression while the other remains in tension. We leave it to the Exercises to show that it is the shorter bar that goes into compression. The displacement δ of the assembly is found by integrating Eq. (3.41).

The problem is different from that of the uniaxial bar in that first the elastic stresses are recovered and subsequently both the stress and the deflection of the assembly change with time. If the deflection had been maintained, the problem would then be one of two bars relaxing separately and each solution could be obtained from Eq. (3.41) with $d\delta/dt = 0$. By letting the displacement vary with time we have coupled the behavior of the two bars.

We solved the relaxation and recovery problems by assuming that various creep laws were applicable to these processes. This was established in Section 2.5 on the basis of experimental results. The reader is cautioned, however, that the validity of this should be checked for the material under analysis. Barring available data, most analysts assume that the creep laws govern relaxation and recovery.

3.5 CLOSURE

It is appropriate at this point to close this chapter by reviewing our findings and emphasizing our terminology.

In a statically determinate structure the stresses can always be obtained from equilibrium whether the loadings are constant or not. In a statically indeterminate structure under constant load there is a redistribution of stress from an initial state to a final stationary state. The latter refers to a state in which stresses are no longer changing. The term "non-stationary" refers to a state during which the stresses are changing. The so-called steady creep approximation states that the creep is governed by a straight line approximation to the creep curve of the material. The terms "steady creep" and "secondary creep" are interchangeable. Moreover, as seen in this chapter the steady creep approximation is used to obtain the stationary solutions, that is, a steady creep law for the *material* is used to find the stationary solution for the *structure* that is under consideration. If the loading is not constant, then a stationary solution cannot develop and the creep is non-stationary. Sometimes the term "transient creep" is used instead of "non-stationary creep."

62 Creep Deformation, Stress Relaxation, and Recovery Problems

We have seen that the steady creep solution was readily obtained for the problems that we considered. If the loads are constant, then this becomes a useful solution. In general, however, this is not an adequate solution because the loadings are seldom constant. Thus a more accurate formulation that governs non-stationary creep, with hardening, must be solved. As we have shown, such solutions cannot usually be obtained in closed form. In the remainder of the book we shall, therefore, discuss the various methods that are used to get around this difficulty.

Finally, we gave illustrations of the fact that a steady creep solution gives a time hardening solution if the time scale is redefined. This forms the basis of the approximate approaches that are presented in the next chapter. In obtaining the steady creep solutions we used the fictitious elastic modulus E_0, whereas in obtaining any solution with hardening we used the actual elastic modulus E.

REFERENCES

[1] S. P. Timoshenko, and J. N. Goodier, *Theory of Elasticity*, 3rd ed. McGraw-Hill Book Co., Inc., New York (1970).
[2] I. Finnie, and W. R. Heller, *Creep of Engineering Materials*, McGraw-Hill Book Co., Inc., New York (1959), p. 52.
[3] J. A. H. Hult, *Creep in Engineering Structures*, Blaisdell Publishing Co., Waltham, Mass., (1966).

EXERCISES

3.1 Formulate the problem of nonstationary creep of a thick, pressurized tube in plane strain with vanishing axial strain.

3.2 Derive the steady creep solution for a rectangular beam under pure bending. What is the curvature as a function of time?

3.3 Derive the steady creep solution of a rectangular beam that is simply supported and subjected to a distributed load.

3.4 Derive the elastic solution and the steady creep solution for the truss shown in Fig. 3.3. What is the vertical deflection history of the vertex of the truss? All bars are pinned and made of the same material.

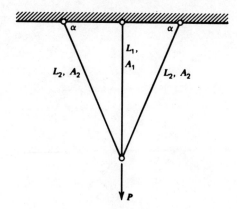

Fig. 3.3 Exercise 3.4.

3.5 Show that the shorter bar goes into compression in the problem of Section 3.4.

CHAPTER 4

APPROXIMATE ANALYTICAL TECHNIQUES

4.1 INTRODUCTION

In the previous chapter we demonstrated the difficulty of general creep and relaxation analysis. As a result of such difficulty several approximate techniques have been proposed and are presented in this chapter. These techniques arrange themselves into two broad categories. First, the reference stress method has been applied to problems of creep deformation under steady and variable loads, creep buckling, and creep rupture. Second, bounding techniques that emanate from the principle of virtual work have been devised for creep deformation under steady and variable loads. Later, in Chapter 8, we discuss the powerful and more widely used numerical methods that have been applied to the creep and relaxation problem.

4.2 THE REFERENCE STRESS METHOD*

The first approximate technique that is presented is the reference stress method (RSM). Although the first reference stress was calculated by Soderberg in the United States in 1941 [1], the method has been under development since the middle 1960's in the United Kingdom. Basically, the

---

*This section is based on the author's review of the reference stress method [32].

REFERENCE STRESS METHOD

Xerox WorkCentre 7845 SMTP Transfer Report

Job Status: **FAILED** Job canceled by user.

Job Information

Device Name:	xusps0651
Submission Date:	28/10/16
Submission Time:	15:21
Images Scanned:	2
Size:	0
Attachment Name:	
Format:	Image-Only PDF
Encrypted E-mail:	No

Message Settings:

Subject:	---
From:	---
Reply To:	---
To:	

idea of the method is that a given structure can be analyzed with data obtained from a single creep test at its reference stress.

In this chapter we discuss the evolution of the RSM and its success in predicting the behavior of typical structures in creep under steady and variable loads, and stress relaxation. In later chapters on creep rupture and creep buckling we apply it to these areas as well.

4.2.1 Development of the Reference Stress Method for Creep under Steady Loads

The aim of the RSM is to correlate the creep deformations in a structure with the results of an equivalent simple creep test. That is to say, the deflection δ at a point in a structure at some time t_0 is given by

$$\delta(t_0) = \beta \varepsilon^C(t_0) \tag{4.1}$$

where β is a geometric scaling factor that depends on the structure and the boundary conditions, $\varepsilon^C(t_0)$ is the creep strain at the time t_0 as obtained in a creep test that is performed on a sample of the material of the structure at the reference stress σ_R. As indicated previously, it appears that Soderberg [1] was the first person to propose this. In 1941 he calculated the reference stress for the correlation of a set of creep tests of pressurized tubes. Some time later Shulte [2] observed that in a solution of creep of beams there were points in the cross section at which the stress did not change as the solution progressed from the initial elastic solution to the final stationary solution at a constant moment. By running a creep test at this constant stress, he was able to make accurate predictions of the beam deflections. His work was formulated for plastics but he surmised that it would apply to metals as well. Subsequently, Anderson et al. [3] analyzed a set of beams with various end conditions for a steady creep law of the form presented in Section 2.6:

$$\frac{d\varepsilon^C}{dt} = k\sigma^m \tag{4.2}$$

where we have used k in place of a as the constant. They discovered that the maximum deflection rate for each beam obeyed an equation of the form

$$\dot{V}_{max} = \beta \dot{\varepsilon}^C(\sigma_R) F(m) \tag{4.3}$$

where β is again a geometric factor, $\dot{\varepsilon}^C(\sigma_R)$ is the creep strain rate at the reference stress σ_R, and F depends only upon the creep exponent m. The form of β, σ_R and F were found to depend on the geometry and end conditions of each beam problem. They identified σ_R and F by "inspection" so as to obtain a function F that varied "very little" with m. That this is plausible is shown in Fig. 4.1 where the distribution of dimensionless bending stress across a rectangular beam under pure bending is shown for various values of m. These curves are typical for a wide variety of structures such as pressurized shells, spinning discs, and so on. It should also be emphasized that the reference stress depends on the characteristic displacement that is to be calculated. Thus in a cantilever beam, for example, the deflection and the rotation of the free end would each have different reference stresses.

The phrase "weakly dependent upon the creep exponent" leads to a variety of possibilities for identification of the reference stress in a given problem. A summary of the approaches that have been used was given by Johnsson [4]. This was recently extended by Boyle [5,6], who demonstrated that the approaches are "stochastically respectable" since they have a probabilistic interpretation. We now illustrate these by applying them to

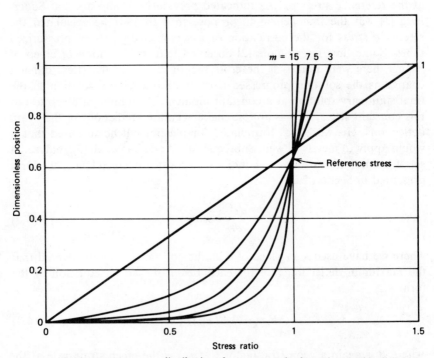

Fig. 4.1 Stationary state stress distributions for a rectangular beam in pure bending [14].

the case of a uniformly loaded cantilever beam that obeys the creep law of Eq. (4.2). The maximum deflection is given by Penny and Marriott [7] as

$$Y_m = \frac{kL^2 t}{(m+1)d} \left(\frac{2m+1}{2m}\right)^m \left(\frac{2\omega L^2}{bd^2}\right)^m \tag{4.4}$$

where d is the depth and b is the width of the rectangular cross-section, L is the length of the beam, and ω is the intensity of the load. The approach of Anderson et al. [3] was to rewrite this expression in the form

$$Y_m = F(m)\frac{L^2}{d}\varepsilon^C(\sigma_R) \tag{4.5}$$

where $\varepsilon^C(\sigma_R)$ is the creep strain at the reference stress

$$\sigma_R = 1.67\left(\frac{\omega L^2}{bd^2}\right) \tag{4.6}$$

and

$$F(m) = \frac{1}{m+1}\left(\frac{6}{5}\right)^m \left(\frac{2m+1}{2m}\right)^m \tag{4.7}$$

As m varies from 3 to 7, $F(m)$ varies from 0.687 to 0.727. The choice of F and σ_R was made, as stated previously, by "inspection."

A more systematic approach to finding the reference stress was proposed by Mackenzie [8]. With the weak dependence of the solution upon m in mind he equated the solutions for m equal to one and the actual value for the material under consideration, M. He therefore found the reference stress to be

$$\sigma_R = 2\alpha\left(\frac{\omega L^2}{bd^2}\right) \tag{4.8}$$

where

$$\alpha = \left(\frac{4}{3(M+1)}\right)^{1/(M-1)} \left(\frac{2M+1}{2M}\right)^{1/(M-1)} \tag{4.9}$$

He found α to have a mean value of 0.79 in the range $1.1 \leqslant M \leqslant 7$ so that

$$\sigma_R = 1.58\left(\frac{\omega L^2}{bd^2}\right) \tag{4.10}$$

A further refinement was proposed by Sim [9, 10], who postulated that the reference stress should be chosen to obtain equal deflections for two arbitrary values of m, say M and M'. With this approach we set $Y_M = Y_{M'}$ to obtain

$$\alpha = \left(\frac{M+1}{M'+1} \left(\frac{2M'+1}{2M'} \right)^{M'} \left(\frac{2M}{2M+1} \right)^{M} \right)^{1/(M'-M)} \tag{4.11}$$

which includes Mackenzie's result (4.9) when $M' = 1$. Sim [9, 10] suggested that the values $M' = 9$ and $M = 3$ give the most reliable results. Thus for the present case we obtain

$$\sigma_R = 1.72 \left(\frac{\omega L^2}{b d^2} \right) \tag{4.12}$$

Finally, Johnsson [4] proposed an additional analytic method for identifying the reference stress. His idea was to find a value of the reference stress that would give a stationary value of the deflection at a given value $m = M$ of the creep exponent; that is,

$$\frac{\partial}{\partial m}(Y_m) = 0 \quad \text{at } m = M \tag{4.13}$$

With this approach we set

$$\frac{\partial}{\partial m}\left(\frac{1}{m+1} \left(\frac{2m+1}{2m\alpha} \right)^m \right) = 0 \tag{4.14}$$

or

$$\frac{\partial}{\partial m}\left[\left(\frac{1}{\alpha} \right)^m f(m) \right] = 0 \tag{4.15}$$

from which Johnsson [4] got the result

$$\alpha(m) = \exp\left(\frac{\partial}{\partial m} \{\ln f(m)\} \right) \quad \text{at } m = M \tag{4.16}$$

so that α depends on the value of m being used. He showed that, for the case of a thick pressurized tube, his definition predicts the deformation with the least error, with Sim's definition, Anderson's definition, and Mackenzie's definition having increasing amounts of error, respectively.

We should emphasize, however, that all of the cited approaches involve procedures that rely on the existence of an analytical solution. This is seldom the case, if ever, in typical problems for which it is more likely that a numerical solution has to be obtained with a digital computer. To handle these situations Sim [9, 10] suggested the following approach, which is presented here for the uniaxial case. Assuming time hardening creep, the uniaxial creep strain rate is related to the uniaxial stress by the equation

$$\frac{d\varepsilon^C}{dt} = Kf(t)\sigma^m \tag{4.17}$$

Now we introduce the dimensionless variables

$$\Sigma = \frac{\sigma}{\sigma_0} \qquad \lambda^C = \frac{\varepsilon^C}{\varepsilon_0} \qquad \varepsilon_0 = \frac{\sigma_0}{E} \tag{4.18}$$

$$\tau = K \int_0^t \frac{f(p)\sigma_0^m \, dp}{\varepsilon_0}$$

with which the creep law takes the form

$$\frac{d\lambda^C}{d\tau} = \Sigma^m \tag{4.19}$$

The extension to multiaxial creep is based on the use of the creep flow equations and the definition of the effective stress and creep strain increment that were given in Section 2.3. Thus the multiaxial counterpart of Eq. (4.19) is

$$\frac{d\lambda_{ij}}{dt} = \tfrac{3}{2}\Sigma_e^{m-1} S_{ij} \tag{4.20}$$

where Σ_e is the dimensionless effective stress and S_{ij} is the stress deviator. On examination of the definition of τ, the dimensionless time, we can write

$$\tau = \frac{\text{creep strain at } t \text{ due to } \sigma_0}{\text{elastic strain at } t \text{ due to } \sigma_0} \tag{4.21}$$

This particular form of dimensionless time was proposed for creep analysis by Penny [11].

Assuming, for example, that the structure under consideration is some type of a pressurized shell, an additional non-dimensional quantity, the deflection, is introduced as

$$U = \frac{u}{a\varepsilon_0} \tag{4.22}$$

where u is the radial displacement and a is the inner radius of the shell. Proceeding further from this a dimensionless rate of radial movement is defined as

$$\frac{dU}{d\tau} = \frac{1}{\varepsilon_0 a} \frac{du}{dt} \frac{dt}{d\tau} \qquad (4.23)$$

which becomes

$$\frac{dU}{d\tau} = \frac{1}{a\sigma_0^m K f(t)} \frac{du}{dt} \qquad (4.24)$$

when the definition of the dimensionless time is used. Now we introduce the reference stress in the form

$$\sigma_R = \alpha \sigma_0 \qquad (4.25)$$

where α is to be determined. Then $dU/d\tau$, the dimensionless rate of movement at the unit stress σ_0 is related to $d\bar{U}/d\bar{\tau}$, the dimensionless rate of movement at the reference stress σ_R by the expression

$$\frac{dU}{d\tau} = \alpha^m \frac{d\bar{U}}{d\bar{\tau}} \qquad (4.26)$$

Passing to the steady state, this leads to

$$\frac{dU_{ss}}{d\tau} = \alpha^m \frac{d\bar{U}_{ss}}{d\bar{\tau}} \qquad (4.27)$$

Generally, $dU_{ss}/d\tau$ will vary with m and σ_0. If it is plotted on semi-log paper as a function of m and σ_0, a set of straight lines will result, as shown in Fig. 4.2. This particular set of curves pertains to a rectangular beam in pure bending and the dimensionless rate of movement in the problem is identified as the dimensionless rate of curvature. The results were given by Sim [9]. The particular value of σ_R which causes zero slope, that is, independence of m, is called the reference stress σ_R. Only two computations are, therefore, needed to find σ_R, as follows. Compute $dU_{ss}/d\tau$ for two values m_1 and m_2 of the creep exponent for an arbitrary value of σ_0. Equation (4.27) thus gives

$$\left.\frac{dU_{ss}}{d\tau}\right|_{m_1} = \alpha^{m_1} \left.\frac{d\bar{U}_{ss}}{d\bar{\tau}}\right|_{m_1} \qquad (4.28)$$

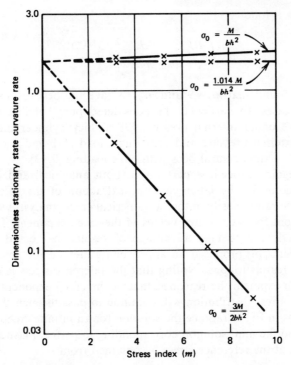

Fig. 4.2 Dimensionless stationary state curvature rates for a rectangular beam in pure bending with various unit stress values (M is the applied moment, b is the width, and h is half of the depth) [9].

and

$$\left.\frac{d\overline{U}_{ss}}{d\bar{\tau}}\right|_{m_2} = \alpha^{m_2} \left.\frac{d\overline{U}_{ss}}{d\bar{\tau}}\right|_{m_2}$$

But the condition for the determination of the reference stress is that the steady state rate of movement that it causes shall be independent of m, that is,

$$\left.\frac{d\overline{U}_{ss}}{d\bar{\tau}}\right|_{m_1} = \left.\frac{d\overline{U}_{ss}}{d\bar{\tau}}\right|_{m_2} \tag{4.29}$$

This, in conjunction with Eq. (4.28), gives

$$\alpha^{-m_1}\left.\frac{dU_{ss}}{d\tau}\right|_{m_1} = \alpha^{-m_2}\left.\frac{dU_{ss}}{d\tau}\right|_{m_2} \tag{4.30}$$

from which it follows

$$\alpha = \left(\frac{dU_{ss}}{d\tau}\bigg|_{m_1} \div \frac{dU_{ss}}{d\tau}\bigg|_{m_2} \right)^{1/(m_1 - m_2)} \quad (4.31)$$

Sim [9, 10] found that $m_1 = 9$ and $m_2 = 3$ gave the best results for a wide variety of cases. In particular, he considered pure bending of beams, a thick sphere under internal pressure [9], a thick cylinder under internal pressure, external pressure and centrifugal load [12], and a spinning disc mounted on a rigid central boss with edge loading [9, 13].

The foregoing procedure pertains to both analytical and digital computer solutions. In the latter case the extraction of the reference stress could still be rather costly in that a numerical creep analysis would have to be carried out for each of two values of the creep exponent. To avoid this type of work and to take advantage of results that could possibly be available, Sim [14] proposed an approximate method for determining the reference stress as follows. Noting that the reference stress is independent of the creep exponent he reasoned that as the creep exponent approaches infinity the stress distribution will continue to pass through the point that defines it, as in Fig. 4.1. Since the solution for an infinite creep exponent is analogous to the limit solution corresponding to perfect plasticity, Sim [14] proposed that the reference stress be obtained from

$$\sigma_R = \frac{P}{P_L} \sigma_y \quad (4.32)$$

where P is the load on the structure, P_L is the limit value of the load, and σ_y is the yield stress. This formula has far-reaching consequences, as we show later in this chapter. To apply it the limit load is assumed to be available, that is, it can be found analytically (see, for example, the text by Hodge [15]), or experimentally from a test of a model of the problem. Moreover, Ponter and Leckie [16] have shown that deformations that are obtained with a reference stress obtained from Eq. (4.32) will constitute an upper bound on the actual deformations. Thus we are provided with a powerful means of performing creep analyses. The accuracy of the approach for some structures for which an analytical limit load is available is shown in Table 4.1. There, reference stresses obtained analytically by Mackenzie [8] are given along with reference stresses obtained with Eq. (4.32).

Once the reference stress has been found by any of the methods described above, we are in a position to determine the creep deformation due to a steady load. First we plot, for the sample case of a pressurized tube, the dimensionless displacement $\bar{U}$ caused at the bore of the tube for

TABLE 4.1 Comparison of Reference Stresses

Structure	σ_R, Ref. [8]	σ_R, Eq. (4.32)
Pure bending of a rectangular beam	$1.014 \dfrac{M}{bd^2}$	$\dfrac{M}{bd^2}$
Uniformly loaded cantilever beam	$1.575 \dfrac{wL^2}{bd^2}$	$2 \dfrac{wL^2}{bd^2}$
Central point load on a simply supported circular plate	$0.534 \dfrac{P}{d^2}$	$0.638 \dfrac{P}{d^2}$
Central point load on a clamped circular plate	$0.339 \dfrac{P}{d^2}$	$0.320 \dfrac{P}{d^2}$

Source: Reference 7.

values of m running from two to nine as a function of the dimensionless time $\bar{\tau}$. The barred variables denote the fact that the reference stress σ_R is being used in defining the dimensionless variables of the problem. This is given in Fig. 4.3, which is typical of all such results obtained by Sim [9, 10]. The figure shows that the steady state deformation rate $d\bar{U}_{ss}/d\bar{\tau}$ is independent of the creep exponent. This fact has already been used in determining the reference stress; now it can be used to determine bounds on the deformation at the bore of the tube, for example, as depicted in general in Fig. 4.4. There we observe that a line with slope $d\bar{U}_{ss}/d\bar{\tau}$ drawn from the initial elastic response at $\bar{\tau}=0$ gives a lower bound, while another line drawn with the same slope from an initial point that is a multiple γ of the initial elastic response gives an upper bound on the deflection. The initial elastic response $\bar{U}_e$ is obtained by an elastic analysis. The quantity γ is found from a steady creep analysis for a large value of m, say nine and by extension of a tangent to that response back to $\bar{\tau}=0$. Therefore, we may write

$$\bar{U}_L = \bar{U}_e + \bar{\tau}\frac{d\bar{U}_{ss}}{d\bar{\tau}} \qquad \text{lower bound}$$

$$\bar{U}_u = \gamma\bar{U}_e + \bar{\tau}\frac{d\bar{U}_{ss}}{d\bar{\tau}} \qquad \text{upper bound}$$

$$= \bar{U}_L + (\gamma - 1)\bar{U}_e \qquad (4.33)$$

74 Approximate Analytical Techniques

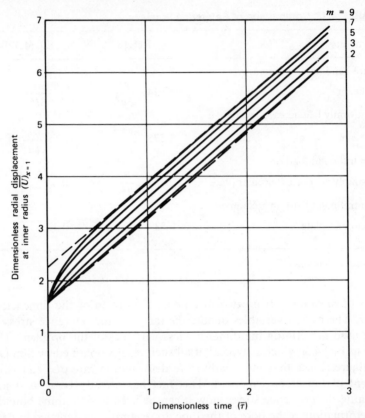

Fig. 4.3 Dimensionless radial displacement at the bore of a thick pressurized tube [12].

Incidentally, from this construction it can also be seen that a reference stress based on the limit load, Eq. (4.32), is an upper bound since it corresponds to a solution for a large value of the creep exponent, namely $m \to \infty$. The steady state deformation rate can be determined analytically. Once the reference stress is known the bounds can be converted into dimensional quantities as follows, using the lower bound equation

$$u = a\bar{\varepsilon}_0 \bar{U}_e + a\bar{\varepsilon}_0 \frac{d\bar{U}_{ss}}{d\bar{\tau}} \bar{\tau} \qquad (4.34)$$

or

$$u = a\left[\bar{U}_e \cdot (\text{elastic strain due to } \sigma_R \text{ at } t) + \frac{d\bar{U}_{ss}}{d\bar{\tau}} \cdot (\text{creep strain due to } \sigma_R \text{ at } t) \right]$$
(4.35)

and similarly for the upper bound.

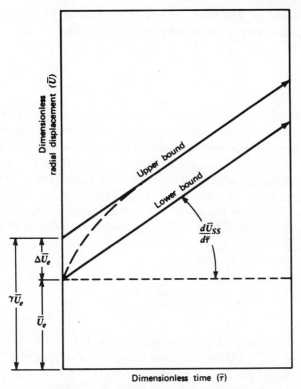

Fig. 4.4 General construction for obtaining bounds on deformations from the reference stress creep solution [12].

The elastic and creep strain at a given time t caused by the reference stress σ_R can be obtained from a single creep test on a tensile specimen of the material. The test is performed at the reference stress. The only assumptions inherent in the process are that the material should obey a creep law of the form Eq. (4.17) and that the creep exponent should be in the range $1 < m \leq 9$. This is usually satisfied by metals. The results of the test can be used directly in Eq. (4.35) and others like it in conjunction with $\bar{U}_e$ from an elastic analysis and $(d\bar{U}_{ss}/d\bar{\tau})$ from a steady creep analysis.

4.2.2 Extension to Combined Loadings

In their study of thick cylindrical tubes under internal pressure, external pressure, and centrifugal loading, Sim and Penny [12] found that the reference parameters for combined loadings could be determined approximately by linear superposition of the reference parameters of the individual loadings. The most important of these is, of course, the reference stress, and this can be found from the formula

$$\bar{\sigma}_{RC} = \sum_{i=1}^{N} \bar{\sigma}_{Ri} \qquad (4.36)$$

where $\bar{\sigma}_{RC}$ is the reference stress for the combined loading, $\bar{\sigma}_{Ri}$ are the reference stresses for the individual loadings, and N is the number of loadings. The reference stress calculated this way by Sim and Penny [12] agreed within 0.2% with the reference stress taken from a combined load solution for plane strain of thick cylinders. For plane stress Sim [13] found that the agreement was not good. Both investigations [12,13] showed, however, that whenever the sense of each loading was radially outward the agreement was excellent. Other reference parameters, discussed previously, are $\bar{U}_e$, $d\bar{U}_{ss}/d\bar{\tau}$, and γ. These are obtained from a combined loading situation by using the formula [12]

$$\bar{p}_c = \sum_{i=1}^{N} \left(\frac{L}{\bar{L}} \bar{p} \right)_i \qquad (4.37)$$

where $\bar{p}_c$ is the combined reference parameter, L is one of the N loads in dimensionless form, $\bar{L}$ is the reference value of that load applied alone, and $\bar{p}$ is the reference parameter associated with $\bar{L}$ when it is applied alone. Charts of the appropriate parameters for individual loadings on cylinders and discs are given in [12,13]. As an example, the dimensionless elastic radial displacement at the reference stress of some point in a cylinder under the influence of internal pressure p_i, external pressure p_e and centrifugal loading $\rho \omega^2 a^2$ would be expressed as

$$\bar{U}_e(\text{due to } \alpha, \beta, h) = \frac{\alpha}{\bar{\alpha}} \bar{U}_e(\text{due to } \bar{\alpha}) +$$

$$\frac{\beta}{\bar{\beta}} \bar{U}_e(\text{due to } \bar{\beta}) + \frac{h}{\bar{h}} \bar{U}_e(\text{due to } \bar{h}) \qquad (4.38)$$

Here $\alpha = p_i/\sigma_0$, $\beta = p_e/\sigma_0$, $h = \rho \omega^2 a^2/\sigma_0$, and the bars denote similar quantities at their reference values. The error associated with results from equations such as Eq. (4.37, 4.38) is around 10% when compared to the results of an actual combined analysis. Again the best accuracy was obtained with these formulas when the sense of all of the loadings was radially outward [12,13]. Inaccuracies in this procedure should not be surprising since Eqs. (4.36, 4.37) represent linear combinations of solutions to a nonlinear problem.

4.2.3 Extension to Thermal Loadings

The previous discussion of reference stress concepts has been limited to isothermal behavior. This will seldom be the case in power plant practice, for example. Therefore, it is now appropriate to consider the effect of the temperature on the previous development. Two persons have considered this problem. In 1973 papers by Sim [17] and Johnsson [4] appeared in which the idea of a reference temperature was proposed.

To include thermal effects in the creep analysis both Sim and Johnsson wrote the uniaxial creep law in the form (compare with Eq. (4.17))

$$\frac{d\varepsilon^C}{dt} = Kf(t)\exp(\gamma T)\sigma^m \qquad (4.39)$$

where γ is a parameter that brings in the dependence on temperature at constant stress. This form is held to be valid for moderate variations of the temperature T. It should be emphasized that, as shown in Eq. (4.39), one of the main effects of temperature is an alteration of the creep properties. The foregoing equation is nondimensionalized as follows:

$$\frac{d\lambda^C}{d\tau} = \exp(\xi\chi)\Sigma^m \qquad (4.40)$$

where

$$\Sigma = \frac{\sigma}{\sigma_0} \qquad \lambda = \frac{\varepsilon^C}{\varepsilon_0} \qquad \varepsilon_0 = \frac{\sigma_0}{E} \qquad \xi = \gamma\Delta T_a$$

$$\chi = \frac{T - T_0}{T_a - T_b} = \frac{\Delta T}{\Delta T_a}$$

$$\tau = K\int_0^t f(p)\exp(\gamma T_0)\sigma_0^m \frac{dp}{\varepsilon_0}$$

$$\tau = \frac{\text{creep strain at } t \text{ due to } \sigma_0 \text{ at } T_0}{\text{elastic strain at } t \text{ due to } \sigma_0 \text{ at } T_0} \qquad (4.41)$$

Here T_0 is a reference temperature and T_a, T_b are the inner and outer surface temperatures in the case of a shell, say, such that $T_a > T_b$. The multiaxial formulation of this law follows in the usual way and is not given here.

Sim [17] and Johnsson [4] each used a different approach to the selection of the reference temperature T_0. Johnsson's is simpler and is described first. In a problem concerning radial heat flow in a thick, pressurized cylinder he arbitrarily chose the temperature at the inner surface of the cylinder to be the reference temperature. He then calculated the reference stress for the combined pressure and thermal stress problem according to his previously described approach, see, for example, Eq. (4.16), as a multiple of the applied pressure.

Sim's approach [17] went the opposite way, that is, he began with the isothermal reference stress for a pressurized tube, for example. Then the stationary displacement rate at the inner radius of a pressurized, heated tube was evaluated. In doing the latter, the location X_r of the reference temperature T_0 was left arbitrary, and the stationary displacement rate at the inner surface was plotted as a function of X_r, as is shown in Fig. 4.5.

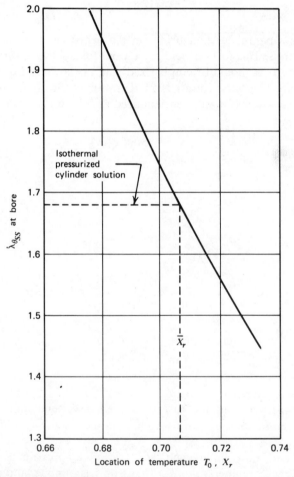

Fig. 4.5 Variation of dimensionless stationary displacement rate at the bore of a thick, pressurized, heated tube as a function of the location of the reference temperature [17].

Now it should be recalled that the *isothermal* reference stress is found as the value that renders the stationary displacement rate at the inner surface of the cylinder independent of the creep exponent m. This would be obtained from a plot that is analogous to the one given in Fig. 4.2 for bending of a beam. Therefore, Sim [17] reasoned that in order to keep the stationary state solution independent of both material properties (m) and loadings, that is, temperature in the present context, the reference temperature should be chosen to be the temperature at the point X_r, which gives the same stationary displacement rate at the inner surface of a heated tube as is obtained at the inner surface of an isothermal tube at the reference stress. Thus in Fig. 4.5 one finds $X_r = 0.706$ for $\dot{\lambda}_{\theta ss}(1) \simeq 1.68$ from the isothermal pressurized cylinder solution.

Sim [17] carried the reasoning one step further on the basis of plots such as we have given in Fig. 4.1. Using these he concluded that the reference stress and reference temperature are located, to within a small error, at the same radial position. Thus, with the isothermal stress distribution and the temperature profile on hand, the necessary quantities can be determined.

4.2.4 A Reference Stress Method for Stress Relaxation

Spence and Hult [18] considered several methods for the analysis of stress relaxation in structures. As illustrative solutions they used the pure bending of a thin tube and a rectangular beam and a pressurized thick sphere. One of the methods they suggested was to use the reference stress σ_R from creep as the initial stress in a uniaxial relaxation test at a fixed strain $\varepsilon_R = \sigma_R / E$. We consider their case of pure bending of a beam and assume that the uniaxial stress-strain law is

$$\frac{d\varepsilon}{dt} = \frac{1}{E}\frac{d\sigma}{dt} + Kf(t)\sigma^m \qquad (4.42)$$

where ε is the total strain and the first term on the right is the elastic strain. If the total strain is held constant, then the exact solution is

$$\frac{\sigma(t)}{\sigma(0)} = \left(1 + (m-1)E[\sigma(0)]^{m-1}\int_0^t Kf(\tau)\,d\tau\right)^{-1/(m-1)} \qquad (4.43)$$

When the stress distribution is integrated across the section of a rectangular beam according to the formula

$$M = \int_A \sigma z\, dz \qquad (4.44)$$

then the foregoing becomes

$$\frac{M(t)}{M(0)} = 3\int_0^1 q^2 \left[1+(m-1)t^*q^{m-1}\right]^{-1/(m-1)} dq \qquad (4.45)$$

where $q = z/H$, the cross section is $2B$ by $2H$,

$$t^* = E\left[\hat{\sigma}(0)\right]^{m-1} \int_0^t Kf(\tau)\, d\tau \qquad (4.46)$$

and $\hat{\sigma}(0)$ is the extreme fiber stress at $t=0$, or

$$\hat{\sigma}(0) = \frac{3M(0)}{4BH^2}$$

Equation (4.45) is evaluated numerically. Spence and Hult [18] went on to obtain approximations to Eq. (4.45) by several methods as follows. The first was based on the Kachanov approximation, in which it is assumed that there is no stress redistribution and that, therefore, the stress at any time is related to the initial stress by a constant. As a counterpart to Eq. (4.45) this procedure led to

$$\frac{M(t)}{M(0)} = \left(1 + \frac{3(m-1)}{m+2} t^*\right)^{-1/(m-1)} \qquad (4.47)$$

with t^* given by Eq. (4.46). A second approach assumed another extreme of behavior by postulating that at any time, complete redistribution has taken place and the stresses are proportional to the stationary stresses. With this procedure they obtained

$$\frac{M(t)}{M(0)} = \left(1 + \frac{(2m+1)^m(m-1)}{(3m)^m} t^*\right)^{-1/(m-1)} \qquad (4.48)$$

with t^* again defined by Eq. (4.46). Finally, if the reference stress obtained by Mackenzie [8] for the rectangular beam, that is,

$$\sigma_R = M/8\alpha BH^2$$

where $\alpha = 0.245$, is used to start the relaxation behavior, then

$$\frac{M(t)}{M(0)} = \left[1 + (m-1)(1.47)^{1-m} t^*\right]^{-1/(m-1)} \qquad (4.49)$$

Notice that m still appears even though reference solutions are aimed at eliminating dependence on it. It is retained by Spence and Hult [18] to emphasize the fact that the RSM result and the stationary state approximation are really the same thing. The results obtained by the various approaches, as plotted by Spence and Hult, are reproduced in Fig. 4.6 for the case of pure bending of a rectangular beam. There it is seen that the accuracy of the approximate methods increases as m decreases and the stationary state-reference stress approach gives a better approximation to the

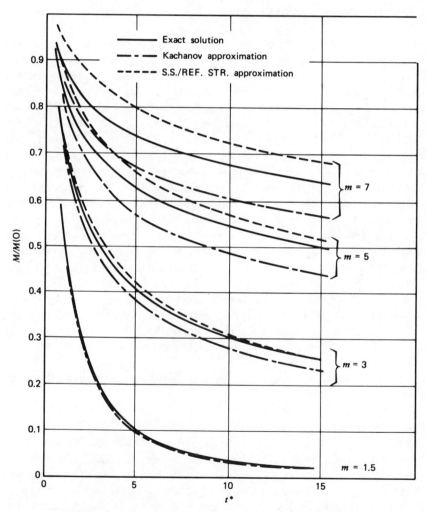

Fig. 4.6 Relaxation of a rectangular beam in pure bending [18].

exact relaxation result than does the Kachanov approach. They also discovered that the accuracy of the RSM improves as the kinematic determinacy increases. The concept was recently applied to curved piping by Boyle and Spence [19]. As an improvement in the analysis of stress relaxation Chern [20] has suggested that since the stress is changing, the creep rate in Eq. (4.42) should be replaced by

$$\dot{\varepsilon}^C = \int_0^t Kf(t-z)m\sigma^m(z)\dot{\sigma}(z)\,dz \qquad (4.50)$$

4.2.5 The Skeletal Point Concept

Marriott and Leckie [21] observed that there are points in components undergoing transient creep at which the stress does not change with time. An example of this is shown in Fig. 4.7, where the effective stress is plotted against radius for a pressurized thick cylinder in creep. As we have mentioned earlier, Schulte [2] was the first to observe this behavior and used it to predict the creep deformation of beams. He did not use the term "skeletal" point to identify the idea. In his review Marriott [22] points out that the stress at the skeletal point is not a reference stress if one keeps in mind the definitions that have been mentioned here. As further pointed out by Marriott [22] confusion has arisen between the reference stress and the stress at the skeletal point because of the sample problems that have been used to illustrate both ideas. Invariably, discussions of the two concepts involved the bending of a beam and the pressurization of a thick tube. These are examples of structures whose strain distributions are independent of the creep law used. That is, the Bernouilli-Euler hypothesis fixes the strain distribution in the beam, while equilibrium can be used to establish the skeletal point as the intersection between the initial elastic solution and the final steady creep solution in the tube. For such kinematically determinate problems the skeletal stress can be shown to be identical to the reference stress. The conclusion is, however, not true in general.

4.2.6 Experimental Verification of the Reference Stress Method for Steady Loads

Several experimental studies that were aimed at determining the validity of the RSM for steady loads have been carried out. Mackenzie [8] considered tests on the pure bending of magnesium beams of rectangular and "I" cross-section and centrally loaded, simply supported brass beams of rectangular cross-section. He calculated the reference stress by his own procedure that we described in conjunction with our Eqs. (4.8, 4.9, 4.10). He

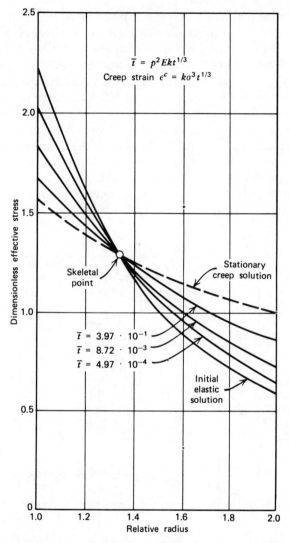

Fig. 4.7 Stress redistribution in thick pressurized tubes [22].

found that the experimental moment-curvature relationships and the load-deflection relationships for the beams involved were predicted very nicely by the RSM. Sim and Penny [14,23] considered the pure bending of aluminum beams. Their results, which are reproduced in Fig. 4.8, show that analyses based upon the RSM do as good a job of predicting the experimental results as do analytical solutions that consider transient effects. Fairbairn [24] tested aluminum tubes in pure bending. He found that the RSM and the time hardening predictions agreed with each other over most of the test period but were each about 50% below the observed creep strain. By increasing the reference stress by only 5% he predicted the observed results very well, as shown in Fig. 4.9. It is, however, unsettling to consider that such a slight change in the reference stress produces such a drastic change in the predictions. In addition, it is to be expected from the development of the RSM that whenever a time hardening analysis doesn't predict the results well the RSM shouldn't either. Boardman et al. [25] measured the extension due to creep in helical springs made of a stainless steel. Although they did not cite the RSM, they showed that in their particular series of tests there was a large amount of transient redistribution and that the ratio of initial to final extension rates was greater than

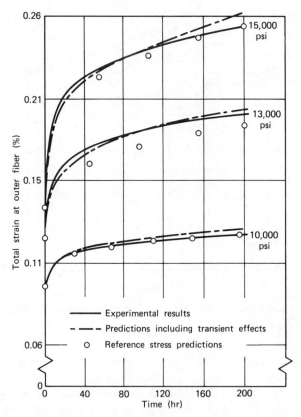

Fig. 4.8 Comparison of theoretical and experimental results for creep of rectangular beams in constant pure bending [23].

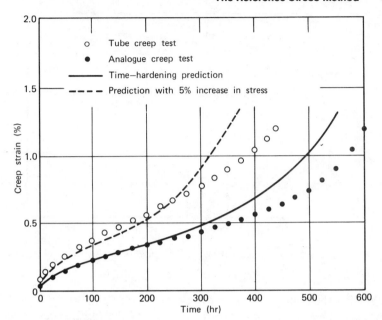

Fig. 4.9 Comparison of theoretical and experimental results for creep of tubular beams in constant pure bending [24].

the value that would have been obtained if the solution were independent of the creep exponent.

Leckie and Ponter [26, 27] investigated aluminum portal frames for which they calculated the reference stress by means of Eq. (4.32). They began by measuring $P_L = 85$ lb for a given frame in a test. Then they selected a load $P = 66$ lb and obtained a displacement history of the creeping portal frame as shown by the solid line in Fig. 4.10. Using Eq. (4.32) with the above values of P_L and P, and a yield stress of 15 ksi, they obtained the reference stress of 11.5 ksi. They then carried out a uniaxial creep test at this stress and by means of the RSM they determined the predicted curve for the portal frame shown by the dotted line in Fig. 4.10. As could be expected the agreement between the results of the RSM and the experimental values improves as time goes on for this constant load case. That the RSM results should lie above the actual results is consistent with the previously stated fact that displacements obtained with reference stresses obtained from Eq. (4.32) were shown to be upper bounds by Ponter and Leckie [16].

Finally, Leckie, Hayhurst, and Morrison [28] considered the creep of a more typical pressure vessel component, namely, an intersection between a

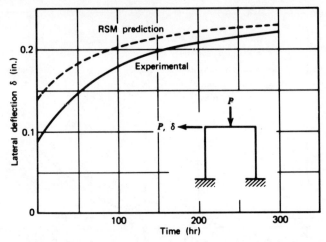

Fig. 4.10 Comparison of theoretical and experimental results for creep of a portal frame under constant load [26].

spherical shell and a cylindrical nozzle under internal pressure. They obtained the required limit pressure for use in Eq. (4.32) several ways. For a given experimental model they calculated a theoretical limit pressure $P_L = 1091$ psi. They also tested the model and found experimental limit pressures of 750 psi (excluding geometric change effects) and 900 psi (including geometric change effects). Then they calculated the reference stresses with Eq. (4.32) for a range of internal pressures. Each of these stresses led to a steady state strain rate from actual creep tests of the model material and then to a theoretical vertical displacement rate of the nozzle. The results are given in Fig. 4.11 in terms of the internal pressure. There we see that the RSM based on the geometrically compatible experimental limit pressure gives the best prediction of the experimental curve.

We may, therefore, conclude that the RSM predicts the behavior of structures under steady loads very well. Given that, it is based on neglecting the stress redistribution, this is not a surprising result. Since redistribution effects are associated with static indeterminacy (see Hult [29]) we should emphasize at this point that the accuracy of the RSM increases as the static determinacy increases. See, for example, References 9 and 18.

4.2.7 Application to Creep Deformation Under Variable Loads

There is no particular benefit to the analysis of a component under steady loads with the RSM. Moreover, it is not surprising that the method should be successful in predicting creep deformations under such loads. In the design of modern components one is far more interested in the analysis of creep under variable loadings and, in particular, cyclic loadings. A com-

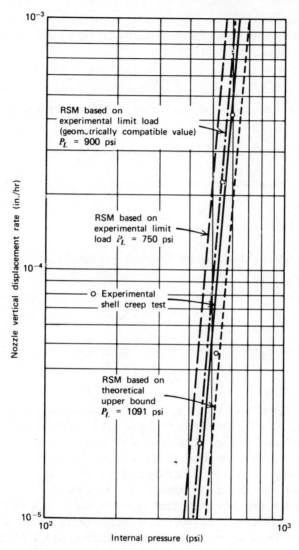

Fig. 4.11 Comparison of theoretical and experimental results for creep of a pressurized cylinder/sphere intersection [24].

prehensive review of the analysis of creep under variable loadings was given by Leckie [27] while a brief summary of some of the approaches to the problem was included in a review by Hayhurst et al. [30].

In this section we present some results that are currently available in the specific area of the use of reference stress ideas in the analysis of creep under variable loadings as proposed by Sim [9].

Sim's development [9] is restricted to variable *proportional* loading of components that obey a creep law given by Eq. (4.17). During proportional variable loading of the form

$$P = P_0 g(t) \tag{4.51}$$

there will be some initial redistribution of stress after which the stresses will vary according to the form

$$\sigma^* = \sigma' g(t) \tag{4.52}$$

The response of a specimen of the material to this stress is, from Eq. (4.17),

$$\frac{d\varepsilon^C}{dt} = Kf(t)\sigma'^m g^m(t) \tag{4.53}$$

We now nondimensionalize the foregoing expression by using the definitions of Eqs. (4.18) to obtain

$$\frac{d\lambda^C}{d\tau} = \Sigma^m \tag{4.54}$$

which is identical to Eq. (4.19) except that the unit stress is defined as

$$\sigma_0 = \sigma_0' g(t) \tag{4.55}$$

and $\Sigma = \sigma'/\sigma_0' = \sigma^*/\sigma_0$, where σ_0 is the unit stress in the variable load problem and σ_0' is the unit stress in the steady load problem.

The significance of Eqs. (4.54) and (4.55) is that the reference stress of the variable load case is equal to the reference stress of the steady load case multiplied by the time variation of the external load Thus for a variable load case the creep response can be estimated from a variable stress creep test that has as its time variation the history of the external load. Sim also alludes to the fact that if the temperature is a variable, then the reference temperature can be found, in analogy to Eq. (4.55), to be

$$T_0 = T_0'' h(t) \tag{4.56}$$

where $h(t)$ is the variation of the temperature in the component and T_0'' is the reference temperature for the steady temperature case.

Sim [14, 23] applied his approach to cyclically loaded beams. The results of these tests are shown in Fig. 4.12 where the outer fiber strains due to a cyclic bending moment are plotted against time for three stress cycles. The figure also shows the results obtained by the RSM in conjunction with the results of a tensile test for each load cycle, that is, Eq. (4.55) and some analytical predictions that take into account the transient (nonstationary) effects. It is seen that the RSM predicted the experimental beam results better than the transient (nonstationary) analysis did. It was argued that this occurred because the RSM includes the effect of the reverse creep as brought into it by the use of a variable stress creep test that is analogous to the actual situation. The transient analysis relies on constant stress data and a hardening hypothesis that does not bring in the actual phenomena associated with variable loading. Furthermore, in a review of Sim's tests Leckie and Ponter [26, 27] proved that the RSM should do so well because very little stress redistribution took place.

Fairbairn [24], whose constant moment tests on tubular beams we cited earlier, also conducted cyclic moment tests. He varied the moments to cycle the extreme fiber stresses between two levels. He also carried out uniaxial tensile tests on the tube material at the same stress cycles for use in conjunction with Eq. (4.55). He too discovered that the RSM gave a better prediction of the experimental creep strains in the tubes than did a transient creep analysis based on constant stress data and a time hardening assumption. Again, this is because the RSM uses the actual cyclic behavior of the material.

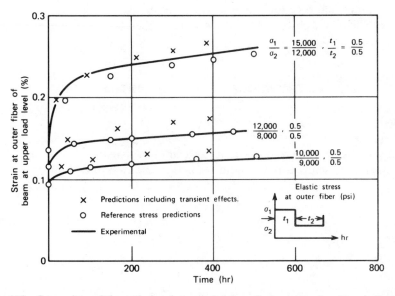

Fig. 4.12 Comparison of theoretical and experimental results for creep of rectangular beams due to cyclic pure bending [14] [23].

4.2.8 Conclusions on the Reference Stress Method

It is evident from the preceding discussion that the RSM has been applied to a wide variety of creep analysis problems. Its success in analyzing these problems has been demonstrated for deformation under steady and variable loads and stress relaxation. Later it is applied to rupture and buckling. In each case it has given satisfactory, conservative estimates of the relevant quantities as demonstrated by its application to the results of the experiments. Throughout the discussion its applicability to the analysis of proportional loading of structures that suffer no redistribution of stress has been emphasized. It is, therefore, appropriate to examine the severity of the restriction to such structures. Ponter and Hayhurst [31] have commented on this. They pointed out that thick tubes under internal pressure, beams under uniform axial load and flexure, bars under torsion, and spinning discs are kinematically determinate. These are simplified structures providing idealizations that are suitable when many more complex structures are under consideration. General multicomponent structures are not expected to be determinate. Nevertheless, it should be emphasized that one is considering an approximate, yet conservative, analysis method. In this regard the RSM has been shown to be quite successful. It is evident from the presentation given here that while the RSM is an approximate method it is definitely *not a simple approach* to creep analysis. Tests must be conducted and certain analyses must still be carried out when the RSM is used. Thus the choice of the RSM will be based largely on the preference of the investigator.

4.3 BOUNDING TECHNIQUES BASED ON VIRTUAL WORK

4.3.1 The Principle of Virtual Work

We start by considering stresses σ_{ij} and strains ε_{ij} in an elastic solid. These stresses and strains can be found from W, the strain energy function, and Ω, the complementary energy function, by means of the formulas [33]

$$\sigma_{ij} = \frac{\partial W}{\partial \varepsilon_{ij}} \qquad \varepsilon_{ij} = \frac{\partial \Omega}{\partial \sigma_{ij}} \qquad (4.57)$$

where

$$W = \int \sigma_{ij} \, d\varepsilon_{ij} \qquad \Omega = \int \varepsilon_{ij} \, d\sigma_{ij} \qquad (4.58)$$

A further restriction is placed on Eqs. (4.57) by Drucker's postulate of material stability. That is, given a pair of strain states ε_{ij} and ε_{ij}^* with corresponding stress states σ_{ij} and σ_{ij}^*, it is required that [34]

$$\int_{\varepsilon_{ij}^*}^{\varepsilon_{ij}} (\sigma_{ij} - \sigma_{ij}^*) \, d\varepsilon_{ij} \geq 0 \tag{4.59}$$

where σ_{ij}^* remains constant during the integration from ε_{ij}^* to ε_{ij} over any path. The foregoing inequality may be written in other forms. If we note that

$$W + \Omega = \int (\sigma_{ij} \, d\varepsilon_{ij} + \varepsilon_{ij} \, d\sigma_{ij}) = \sigma_{ij} \varepsilon_{ij} \tag{4.60}$$

we can also write

$$\Omega(\sigma_{ij}^*) + W(\varepsilon_{ij}) \geq \sigma_{ij}^* \varepsilon_{ij} \tag{4.61}$$

Now consider a continuum of stable elastic material. Body forces F_i act on the continuum. We shall assume that all displacements are small so that geometry changes can be ignored. We suppose that a stress field σ_{ij}^* and a strain field ε_{ij} are known for the continuum. The stress field must be in internal equilibrium with the body forces, that is,

$$\frac{\partial \sigma_{ij}^*}{\partial x_j} + F_i^* = 0 \tag{4.62}$$

Surface transactions T_i^* are defined by the requirements of external equilibrium. Thus at any surface of the continuum

$$\sigma_{ij}^* \nu_j = T_i^* \tag{4.63}$$

where ν_j is the unit outward normal at a point. The strains ε_{ij} must be compatible with the displacements u_i. Thus

$$\varepsilon_{ij}^C = \frac{1}{2}\left(\frac{\partial u_i^C}{\partial x_j} + \frac{\partial u_j^C}{\partial x_i}\right) \tag{4.64}$$

We note that σ_{ij}^*, ε_{ij}^C are completely independent of each other. Since T_i^*, F_i^*, and σ_{ij}^* are in equilibrium and u_i^C and ε_{ij}^C are compatible we may write

92 Approximate Analytical Techniques

by the principle of virtual work [35]

$$\int_A T_i^* u_i^C \, dA + \int_V F_i^* u_i^C \, dV = \int_V \sigma_{ij}^* \varepsilon_{ij}^C \, dV \qquad (4.65)$$

where A and V are the area and volume of the continuum. Now reconsider inequality (4.61). σ_{ij}^* and ε_{ij}^C is an admissible pair of states at all points of the body. Thus we may integrate Eq. (4.61) over the volume, retaining the inequality. Then by substituting Eq. (4.65) into the result, we obtain

$$\int_V \Omega(\sigma_{ij}^*) \, dV + \int_V W(\varepsilon_{ij}^C) \, dV \geq \int_A T_i^* u_i^C \, dA + \int_V F_i^* u_i^C \, dV \qquad (4.66)$$

We note that the equality holds when $\varepsilon_{ij}^C = \varepsilon_{ij}^*$, that is, when σ_{ij}^*, ε_{ij}^C are the solution of some problem.

4.3.2 Application to Creep Problems

Inequality (4.66) was derived by Martin [36]. It was applied by him [37] to steady creep problems on the basis of the elastic analogy cited in Section 2.6. That is a problem governed by the stady creep law

$$\left(\frac{\dot{\varepsilon}}{\varepsilon_0}\right) = \left(\frac{\sigma}{\sigma_0}\right)^m \qquad (4.67)$$

may be transformed into an analogous problem in nonlinear elasticity by considering strain rates as strains, and displacement rates as displacements. For an equation such as Eq. (4.67) the energies can be written as

$$W(\dot{\varepsilon}_{ij}) = \int_0^{\dot{\varepsilon}_{ij}} \sigma_{ij} \, d\dot{\varepsilon}_{ij} = \frac{m}{m+1} \sigma_{ij} \dot{\varepsilon}_{ij}$$

$$\Omega(\sigma_{ij}) = \int_0^{\sigma_{ij}} \dot{\varepsilon}_{ij} \, d\sigma_{ij} = \frac{1}{m+1} \sigma_{ij} \dot{\varepsilon}_{ij} \qquad (4.68)$$

Now substitute Eq. (4.68) into Eq. (4.66) to obtain

$$\frac{1}{m+1} \int_V \sigma_{ij}^* \dot{\varepsilon}_{ij}^* \, dV + \frac{m}{m+1} \int_V \sigma_{ij}^C \dot{\varepsilon}_{ij}^C \, dV \geq \int_A T_i^* u_i^C \, dA \qquad (4.69)$$

where the body force term has been ignored. Here again, σ_{ij}^* is in equilibrium with T_i^* and ε_{ij}^C is compatible with u_i^C. ε_{ij}^* is the strain associated with σ_{ij}^* and need not be compatible, while σ_{ij}^C is associated with ε_{ij}^C and need not satisfy equilibrium.

Now consider the following boundary value problem. Let surface tractions T_i be given on part of the surface A_T and let velocities $\dot{u}_i$ be zero on the remainder of the surface A_u. Suppose that σ_{ij}, $\dot{\varepsilon}_{ij}$ and $\dot{u}_i$ are the solution of the problem. Certainly $\dot{\varepsilon}_{ij}$ satisfies the conditions required on ε_{ij}^C in Eq. (4.69). If we equate the internal and external energy dissipation rates

$$\int_A T_i^C \dot{u}_i^C \, dA = \int_V \sigma_{ij}^C \dot{\varepsilon}_{ij}^C \, dV \tag{4.70}$$

Substitute this into Eq. (4.69) to obtain

$$\int_A \left(T_i^* - \frac{m}{m+1} T_i^C \right) \dot{u}_i^C \, dA \leq \frac{1}{m+1} \int_V \sigma_{ij}^* \dot{\varepsilon}_{ij}^* \, dV \tag{4.71}$$

We now choose the T_i^* system to eliminate the unwanted $\dot{u}_i$ values. On A_T let

$$T_i^* = \frac{m}{m+1} T_i^C + P_i \tag{4.72}$$

where P_i is a point load acting on the surface where a displacement rate bound is required. Let the displacement rate at this chosen point be $\dot{u}_i^*$. On A_u, T_i^* must be chosen so that the body is in external equilibrium with T_i^*. With these choices

$$P_i \dot{u}_i^* \leq \frac{1}{m+1} \int \sigma_{ij}^* \dot{\varepsilon}_{ij}^* \, dV \tag{4.73}$$

This permits the bound to be computed in terms of σ_{ij}^* only.

As an illustration Martin [37] applied his bounding technique to the creep of a cantilever beam subjected to a distributed load 2ω/unit length as shown in Fig. 4.13(a). Suppose we wish to find a bound on the tip displacement rate $\dot{\delta}_t$. The equilibrium solution pertains to the loading in Fig. 4.13(a), where P is a load placed in line with the required deflection rate. The internal moment M^* that is in equilibrium with the forces in Fig. 4.13(b) is

$$M^* = Px + \frac{m}{m+1} \frac{2\omega x^2}{2} \tag{4.74}$$

If the moment curvature relationship in steady creep is

$$\frac{\dot{K}}{\dot{K}_0} = \left(\frac{M}{M_0} \right)^m \tag{4.75}$$

94 Approximate Analytical Techniques

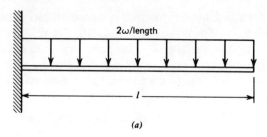

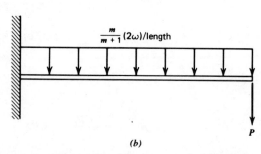

Fig. 4.13 (*a*) The uniformly loaded cantilever beam problem. (*b*) The equivalent load for obtaining a bound on the tip deflection rate.

where $\dot{K}_0$ and M_0 are constants, then the inequality Eq. (4.73) reduces to

$$P\dot{\delta}_t \leq \frac{1}{m+1} \int_0^l M^* \dot{K} \, dx = \frac{1}{m+1} \int_0^l \frac{\dot{K}_0}{M_0^m} (M^*)^{m+1} \, dx$$

Therefore,

$$\dot{\delta}_t \leq \frac{1}{P(m+1)} \int_0^l \frac{\dot{K}_0}{M_0^m} x^{m+1} \left(P + \frac{1}{m+1} \omega x \right)^{m+1} dx \qquad (4.76)$$

For any value of m the integral in Eq. (4.76) can be evaluated, and any value of P can be substituted into the resulting expression. It is also desirable to optimize the right hand side of Eq. (4.76) to obtain the best bound. The results are as follows, for various values of m:

m	$P/\omega l$	Upper Bound	Exact Result
1	0.387	0.254	0.250
3	0.25	0.126	0.125
5	0.15	0.083	0.083

The second column is the optimum value of $P/\omega l$, the third and fourth columns are values of the parameter $M_0^m \delta_t / \dot{K}_0 \omega^m l^{2(m+1)}$. The bound is very good for this simple example.

A lower bound on the solution was developed, with similar reasoning, by Palmer [38]. The methods based on virtual work were extended to other situations such as variable loads, and to interaction with plasticity, by Leckie, Ponter, and others. An excellent review of this effort was given by Leckie [27].

4.4 CLOSURE

In this chapter we have presented two approximate analytical techniques for solving creep problems: the reference stress method and bounding techniques based on the principle of virtual work. It has in fact been shown [16] that deformations found by the reference stress method are also upper bounds. These methods have been used mostly in the United Kingdom. They have not found wide use in the United States where engineers seem to prefer to work with the digital computer and solutions of the type presented in Chapter 8. It can be said, therefore, that both avenues, the approximate analytical one presented here and the numerical one presented later, are now available to those contemplating work in this field. A clear choice has not yet emerged. Approximate analytical methods are presumably simpler, but digital computer solutions are more accurate. On the surface it would seem that the latter would be the preferred method. However, digital computer solutions are becoming more and more expensive so that the simpler approximate methods are becoming more appealing on the basis of cost. It is likely that eventually preliminary design will be carried out with the approximate methods while final design analysis will continue to be carried out with the digital computer.

REFERENCES

[1] C. R. Soderberg, Interpretation of Creep Tests on Tubes, *Trans. A. S. M. E.*, vol. 63, pp. 737–748 (1941).

[2] C. A. Schulte, Predicting Creep Deflections of Plastic Beams, *Proc. A. S. T. M.*, vol. 60, pp. 895–904 (1960).

[3] R. G. Anderson, L. R. T. Gardner, and W. R. Hodgkins, Deformation of Uniformly Loaded Beams Obeying Complex Creep Laws, *J. Mech. Eng. Sci.*, vol. 5, pp. 238–244 (1963).

[4] A. Johnsson, "An Alternative Definition of Reference for Creep," *Int. Con. on Creep and Fatigue in Elevated Temperature Applications*, Inst. Mech. Engrs., Conference Publication No. 13, Paper C205/73, Philadelphia, 1973.

[5] J. T. Boyle, Approximations in the Reference Stress Method for Creep Design, *Mech. Res. Comm.*, (to appear).

[6] J. T. Boyle, A Consistent Interpretation of the Reference Stress Method in Creep Design, *Int. J. Mech. Sci.*, (submitted).

[7] R. K. Penny, and D. L. Marriott, *Design for Creep*, Ch. 4 McGraw-Hill Book Co., Ltd., London, (1971).

[8] A. C. Mackenzie, On the Use of a Single Uniaxial Test to Estimate Deformation Rates in Some Structures Undergoing Creep, *Int. J. Mech. Sci.*, vol. 10, pp. 441–443 (1968).

[9] R. G. Sim, Reference Stress Concepts in the Analysis of Structures During Creep, *Int. J. Mech. Sci.*, vol. 12, pp. 561–573 (1970).

[10] R. G. Sim, Evaluation of Reference Stress Parameters for Structures Subject to Creep, *J. Mech. Eng. Sci.*, vol. 13, pp. 47–50 (1971).

[11] R. K. Penny, Creep of Spherical Shells Containing Discontinuities, *Int. J. Mech. Sci.*, vol. 9, pp. 373–388 (1967).

[12] R. G. Sim, and R. K. Penny, Plane Strain Creep Behavior of Thick-Walled Cylinders, *Int. J. Mech. Sci.*, vol. 13, pp. 987–1009 (1971).

[13] R. G. Sim, Reference Results for Plane Stress Creep Behavior, *J. Mech. Eng. Sci.*, vol. 14, pp. 404–410 (1972).

[14] R. G. Sim, Creep of Structures, Ph.D. Dissertation, University of Cambridge (1968).

[15] P. G. Hodge Jr., *Limit Analysis of Rotationally Symmetric Plates and Shells*, Prentice-Hall, Inc., Englewood Cliffs, N. J. (1963).

[16] A. R. S. Ponter, and F. A. Leckie, Application of Energy Theorems to Bodies Which Creep in the Plastic Range, *J. Appl. Mech.*, vol. 37, pp. 753–758 (1970).

[17] R. G. Sim, Reference Stresses and Temperatures for Cylinders and Spheres Under Internal Pressure with a Steady Heat Flow in the Radial Direction, *Int. J. Mech. Sci.*, vol. 15, pp. 211–220 (1973).

[18] J. Spence, and J. Hult, Simple Approximations for Creep Relaxation, *Int. J. Mech. Sci.*, vol. 15, pp. 741–755 (1973).

[19] J. T. Boyle and J. A. Spence, A Comparison of Approximate Methods for Estimation of the Creep Relaxation of a Curved Pipe, *Int. J. Press. Ves. & Piping*, vol. 4, pp. 221–252 (1976).

[20] J. M. Chern, personal correspondence, October 6, 1976.

[21] D. L. Marriott, and F. A. Leckie, Some Observations on the Deflections of Structures During Creep, *Proc. I. Mech. E.*, vol. 178, Pt. 3L, pp. 115–125 (1963–64).

[22] D. L. Marriott, A Review of Reference Stress Methods for Estimating Creep Deformation, *Proc. IUTAM Symposium on Creep of Structures*, Gothenburg, pp. 137–152, (1970).

[23] R. G. Sim, and R. K. Penny, Some Results of Testing Simple Structures Under Constant and Variable Loading During Creep, *Exptl. Mech.*, pp. 152–159, April 1970.

[24] J. Fairbairn, A Reference Stress Approach to Creep Bending of Straight Tubes, *J. Mech. Eng. Sci.*, vol. 16, pp. 125–138 (1974).

[25] F. D. Boardman, F. P. Ellen, and J. A. Williamson, Measurements of Creep at High Temperatures Using Helical Springs, *J. Strain Analysis*, vol. 1, pp. 140–144 (1966).

[26] F. A. Leckie, and A. R. S. Ponter, Theoretical and Experimental Investigation of the Relationship Between Plastic and Creep Deformation of Structures, *Arch. Mech.*, vol. 24, pp. 419–437 (1972).

[27] F. A. Leckie, A Review of Bounding Techniques in Shakedown and Ratchetting at Elevated Temperatures, *Welding Research Council Bulletin No.* 195, pp. 1–32 (1974).

[28] F. A. Leckie, D. R. Hayhurst, and C. J. Morrison, Creep Deformations of a Shell Intersection Subjected to Constant and Variable Pressure, University of Leicester, Engineering Department Report 74-11, July, 1974.

[29] J. Hult, *Creep in Engineering Structures*, Ch. 6, Blaisdell Publishing Co., Waltham, Mass. (1966).

[30] D. R. Hayhurst, D. A. Kelly, F. A. Leckie, C. J. Morrison, A. R. S. Ponter, and J. J. Williams, "Approximate Design Methods for Creeping Structures," *Int. Conf. on Creep and Fatigue in Elevated Temperature Applications*, Inst. Mech. Engrs., Conference Publication No. 13, Paper C219/73, Philadelphia, 1973.

[31] A. R. S. Ponter, and D. R. Hayhurst, Lower bound on the Time to Initial Rupture of Creeping Structures, *J. Mech. Eng. Sci.*, vol. 15, pp. 357–364 (1973).

[32] H. Kraus, Reference Stress Concepts for Creep Analysis, *Welding Research Council Bulletin No.* 227, June 1977.

[33] I. S. Sokolnikoff, *Mathematical Theory of Elasticity*, McGraw-Hill Book Co., Inc., 2nd ed., New York (1956).

[34] D. C. Drucker, *Proc. 1st U.S. Nat. Congr. Appl. Mech.*, p. 487 (1951).

[35] W. T. Koiter, General Theorems for Elastic Plastic Solids, in I. Sneddon, and R. Hill, eds., *Progress in Solid Mechanics*, vol. I, Ch. IV, 167–221 (1960).

[36] J. B. Martin, A Displacement Bound Technique for Elastic Continua Subjected to a Certain Class of Dynamic Loading, *J. Math. Phys. Solids*, vol. 12, pp. 165–175 (1964).

[37] J. B. Martin, A Note on the Determination of an Upper Bound on Displacement Rates for Steady Creep Problems, *J. Appl. Mech.*, vol. 33, pp. 216–217 (1966).

[38] A. C. Palmer, A Lower Bound on Displacement Rates in Steady Creep, *J. Appl. Mech.*, vol. 34, pp. 216–217 (1967).

CHAPTER **5**

CREEP RUPTURE

5.1 INTRODUCTION

Up to this point our discussions have been limited to creep deformation. Thus they have been aimed at presenting information to permit design against loss of function due to excessive deformation. Now we turn our attention to a failure mode that involves actual breakage of a component. That is, creep rupture or as some prefer to call it, stress rupture. In this case we are trying to understand the actual time dependent rupture of parts. In our previous discussion mathematical models and analyses have been concerned with primary and secondary creep. We will not consider the final tertiary state of the creep process. Because of the unstable nature of tertiary creep our concern will be with rupture itself.

The discussion takes place in three parts. First we discuss creep rupture data, its extrapolation, and its use in the analysis of rupture. Then we will present several analyses of the overall process of creep rupture, in which we bring in a new variable that is known as damage. Finally, given the complexity of the overall analysis of the creep rupture of a structure, we present one approximate method for analyzing it. This is based on the reference stress idea that was developed for analyzing creep deformation in Chapter 4.

5.2 CREEP RUPTURE DATA

Several methods of presenting basic creep rupture data exist. We have seen one of these in Chapter 1, where Fig. 1.5 showed a typical set of curves in which stress was plotted against rupture time for several temperatures. It is not, however, appropriate to design right up to rupture. Thus some type of safety factor has to be introduced in order to avoid failure. In creep this is often handled by replacing the rupture curve with curves that represent the time to reach a certain total strain at a given stress, for example, 0.1%, 0.2%, or 0.5% strain. There are two other ways of presenting creep rupture data. One method plots the stress to cause actual rupture in a given time period against temperature. The time period chosen is usually 100,000 or 200,000 hours. If we note that there are 8760 hours in a year, this means either 11.4 or 22.8 years. This method of plotting refers to creep strength as the stress required to rupture a specimen in the given time period. The other method refers to creep resistance. This is the stress required to achieve a certain strain; again, 0.1%, 0.2%, or 0.5% in the given time period of 100,000 or 200,000 hours. The choice of method depends on the user.

To collect data in support of the foregoing methods of presentation would require, in many instances, testing periods that are of the same order of length as the life of the equipment that is being designed. Since it is not practical to run materials studies that take so long before the design and construction process can even begin, methods have evolved to permit short term tests to be extrapolated to the longer terms that are involved in the design process. In the end one wants to know if the part being designed is or isn't going to rupture within its intended design life. We will now consider how this can be done [1].

Usually, families of rupture curves appear as they do in Fig. 5.1, which is similar to Fig. 1.5. If a temperature-compensated time axis could be derived, one would have what is known as a master curve, shown in Fig. 5.2. Here the point A corresponds to a stress σ, and has a rupture time of 1000 hours at T_1, and a rupture time of 100,000 hours at T_2. The existence of the extrapolation depends on finding a function that has the form

$$f(\sigma, t_R, T) = \text{constant} \tag{5.1}$$

More typically it is written as

$$f(t_R, T) = P(\sigma) \tag{5.2}$$

This would express the idea proposed in Fig. 5.2. Now we examine several cases of this equation.

100 Creep Rupture

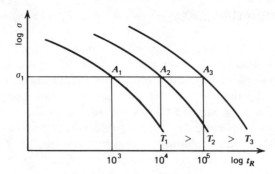

Fig. 5.1 Creep rupture curves.

5.2.1 Larson-Miller Parameter

It is assumed that creep is a rate process governed by the Arrhenius equation

$$\frac{d\varepsilon^C}{dt} = A \exp\left(-\frac{\Delta H}{RT}\right) \qquad (5.3)$$

Here A is a constant, ΔH is the activation energy for creep processes, R is the gas constant, and T is the temperature. On the basis of experiment it is also assumed that the minimum creep rate satisfies

$$t_R \left(\frac{d\varepsilon^C}{dt}\right)_{min} = \text{constant} \qquad (5.4)$$

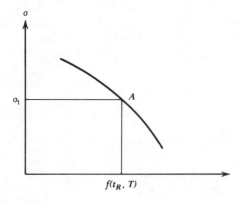

Fig. 5.2 Creep curves merged with temperature compensated abscissa.

If Eqs. (5.3) and (5.4) are combined we get

$$At_R \exp\left(-\frac{\Delta H}{RT}\right) = \text{constant} \tag{5.5}$$

Now if it is assumed that ΔH is a function of stress only, taking the logarithms of both sides gives

$$\log_{10} t_R + \log_{10}\left(\exp -\frac{\Delta H}{RT}\right) = \log_{10} C' \tag{5.6}$$

where C' is the constant in Eqs. (5.4) and (5.5). Since

$$\log_{10} x = \log_e x \log_{10} e \tag{5.7}$$

Eq. (5.7) becomes

$$\log_{10} t_R - 0.434 \frac{\Delta H}{RT} = \log_{10} C' = C$$

or

$$T(\log_{10} t_R + C) = 0.434 \frac{\Delta H}{R}$$

Finally, we write the equation in the form

$$T(C + \log_{10} t_R) = P_{LM}(\sigma) \tag{5.8}$$

where $P_{LM}(\sigma)$ is the Larson-Miller extrapolation parameter [2]. It depends only on stress. The constant $C = \log_{10} C'$ is found by plotting lines of $\log_{10} t_R$ versus $1/T$ for constant stress, as in Fig. 5.3. The resulting lines converge at

$$\frac{1}{T} = 0, \quad \log_{10} t_R = -C$$

Larson and Miller picked $C = 20$ for most materials so that the final expression for the extrapolation equation is

$$P_{LM}(\sigma) = T(20 + \log_{10} t_R) \tag{5.9}$$

A typical curve is given in Fig. 5.4.

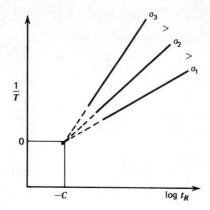

Fig. 5.3 Larson-Miller construction [2].

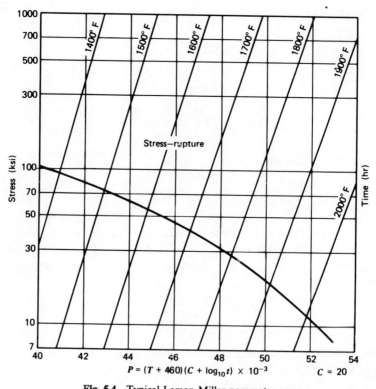

Fig. 5.4 Typical Larson-Miller parameter curve.

5.2.2 Manson-Haferd Parameter

Many investigators questioned the use of $C=20$ for all materials, so several other extrapolation parameters were proposed. One example is the Manson-Haferd parameter [3]. They plotted $\log_{10} t_R$ as a function of temperature, as in Fig. 5.5. The equation for each curve is

$$T_a - T = -f(\sigma)(\log_{10} t_R - \log t_a) \tag{5.10}$$

where (T_a, t_a) is the convergence point of the curves. This can be rewritten as

$$\log_{10} t_R - \log_{10} t_a = \frac{T - T_a}{f(\sigma)}$$

Finally,

$$P_{MH}(\sigma) = \frac{T - T_a}{\log_{10} t_R - \log_{10} t_a} \tag{5.11}$$

is the Manson-Haferd parameter. In this case Larson-Miller's $C=20$ has been replaced by the somewhat more general (t_a, T_a) for each material.

All of the parameters that have been proposed can be expressed as Manson's generalized parameter [4]:

$$P(\sigma) = \frac{\sigma^v \log_{10} t_R - \log_{10} t_a}{(T - T_a)^w} \tag{5.12}$$

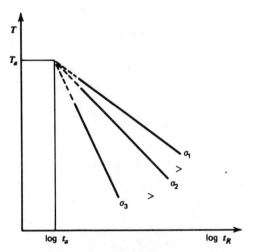

Fig. 5.5 Manson-Haferd construction [3].

If $\nu=0$, $w=-1$, $T_a=0$ and $-\log_{10} t_a = 20$ we obtain the Larson-Miller parameter. If $\nu=0$, $w=1$, $T_a \neq 0$ we get the Manson-Haferd parameter.

5.3 ANALYSIS OF THE RUPTURE PROCESS

Up to now we have studied the general nature of creep rupture (resistance) curves and we have seen how the data are extrapolated. Now we wish to build an analytical model of the rupture process. To begin, we discuss the concept of damage [5].

5.3.1 The Damage Concept

We define the damage in terms of net area A_R of a cross-section that remains to carry the load in a member as a result of some internal flaw distribution

$$D \equiv 1 - \frac{A_R}{A_0} \qquad 0 \leqslant D \leqslant 1 \tag{5.13}$$

Here A_0 is the initial area and A_R is the area after some damage has occurred. We note that the damage goes from zero to unity and, for the present, that it occurs because of creep. A similar development can be carried out for fatigue, where time is replaced by cycles of the load. To continue, we let the momentary stress be σ_R and the initial stress be σ_0. Thus,

$$A_0 D = A_0 - A_R$$

or

$$A_0(1-D) = A_R \tag{5.14}$$

The initial stress in terms of the load is

$$\sigma_0 = \frac{F}{A_0} \tag{5.15}$$

while the current stress is

$$\sigma_R = \frac{F}{A_R} = \frac{\sigma_0 A_0}{A_R} \tag{5.16}$$

Since

$$A_R \sigma_R = A_0 \sigma_0 \qquad (5.17)$$

$$\sigma_R = \frac{\sigma_0}{1-D} \qquad (5.18)$$

$$D = 1 - \frac{\sigma_0}{\sigma_R} \qquad (5.19)$$

The rate of damage is assumed to be

$$\frac{dD}{dt} = f_1(\sigma_R)$$

In particular, we take a power law form so that

$$\frac{dD}{dt} = C_1 \left(\frac{\sigma_R}{\sigma^*}\right)^n \qquad (5.20)$$

where C_1, σ^*, and n are parameters. Incidentally, if we were developing a model for fatigue damage we would, at this point, write

$$\frac{dD}{dN} = C_2 \left(\frac{\sigma_R}{\bar{\sigma}}\right)^m \qquad (5.21)$$

where N is now the cycle and C_2, $\bar{\sigma}$, and m are parameters. The remainder of the derivation is similar for creep and fatigue and will be carried out here for creep.

We solve Eq. (5.20) under the conditions that $D(0)=0$ and $D(t_R)=1.0$ where t_R is the rupture time. Before doing this we substitute Eq. (5.18) into Eq. (5.20) to obtain

$$\frac{dD}{dt} = \frac{C_1}{\sigma^{*n}} \left(\frac{\sigma_0}{1-D}\right)^n \qquad (5.22)$$

Now rearrange the equation to the form

$$(1-D)^n \, dD = C_1 \left(\frac{\sigma_0}{\sigma^*}\right)^n dt$$

apply integration limits

$$\int_0^1 (1-D)^n \, dD = C_1 \left(\frac{\sigma_0}{\sigma^*}\right)^n \int_0^{t_R} dt \qquad (5.23)$$

and integrate to obtain

$$t_R = \frac{1}{C_1} \frac{1}{n+1} \left(\frac{\sigma^*}{\sigma_0}\right)^n \tag{5.24}$$

If we take the logarithm of both sides, we get

$$\log_{10} t_R = \log_{10}\left(\frac{\sigma^{*n}}{C_1(n+1)}\right) - n \log_{10} \sigma_0$$

which has the general form

$$\log_{10} t_R = \text{constant} - n \log_{10} \sigma_0$$

This has the correct experimentally observed form, that is, a curve on log-log coordinates that slopes down to the right. Now rearrange Eq. (5.24) to the form

$$C_1 \left(\frac{\sigma_0}{\sigma^*}\right)^n = \frac{1}{t_R(n+1)} \tag{5.25}$$

and substitute this into the integral equation form Eq. (5.23)

$$\int_0^{t_R} \frac{dt}{t_R} \frac{1}{n+1} = \int_0^1 (1-D)^n \, dD$$

This leads on integration to

$$\int_0^{t_R} \frac{dt}{t_R} = 1 \tag{5.26}$$

for creep damage. For fatigue damage it would lead to the expression

$$\int_0^{N_F} \frac{dN}{N_F} = 1 \tag{5.27}$$

where N_F is the number of cycles to failure.

5.3.2 Multiple Stress Levels

The preceding development pertained to damage due to a single stress level σ_0. For a sequence of constant stresses the integrals in Eqs. (5.26) and

(5.27) can be replaced by sums. Thus Eq. (5.26) leads to

$$\sum_{k=1}^{K} \frac{\theta_k}{t_{Rk}} = 1 \qquad (5.28)$$

for failure, where θ_k is the time spent at a given stress level, and t_{Rk} is the rupture time for that stress. Similarly, Eq. (5.27) leads to

$$\sum_{l=1}^{L} \frac{\nu_l}{N_{Fl}} = 1 \qquad (5.29)$$

for failure, where ν_l is the number of cycles at a given stress and N_{Fl} is the life at that stress in fatigue. The first rule is usually referred to as Robinson's Rule while the second rule is commonly referred to as Miner's Rule. Both rules signify the fact that either in creep or in fatigue the life is used up fractionally by every stress level that exists in a loading history. Moreover, if cycling occurs at elevated temperature, we have the possibility that both types of damage occur simultaneously. Hence, as a first approximation we assume that the total damage must be less than unity. Thus

$$\sum_{k=1}^{K} \frac{\theta_k}{t_{Rk}} + \sum_{l=1}^{L} \frac{\nu_l}{N_{Fl}} = 1 \qquad (5.30)$$

for failure. This assumes that there is no interaction between the fatigue and creep processes. That is, they do not influence each other. This is, therefore, a linear interaction. In Chapter 9 we return to this point as we study the actual interaction. When considering multiple stress levels for creep rupture and the use of Eq. (5.28), two possibilities arise. The stress level may actually fluctuate in a stepwise fashion as sketched in Fig. 5.6(a) or the stress may vary continuously as in Fig. 5.6(b). Eq. (5.28) can be used directly in situations such as Fig. 5.6(a). It can be used for Fig. 5.6(b) if the history is broken up into discrete steps, as shown.

A limitation of the foregoing development is that the nature of the stress field has not been specified. We have implied that it is uniaxial but we have not mentioned anything about whether or not it is uniform across the section. If we adopt the maximum tensile principal stress criterion of failure, then we can apply the foregoing approach to all types of situations as long as we always apply the reasoning to the greatest tensile principal stress in the cross section. However, the stress has to be obtained by a creep analysis. As we saw in Section 3.2.2 in a thick pressurized tube, for

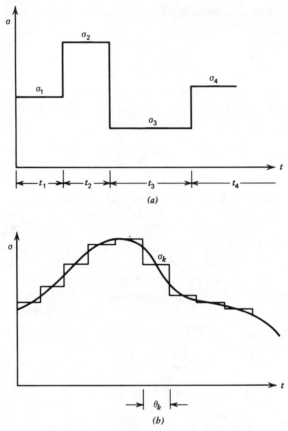

Fig. 5.6 (*a*) Stress history made up of discrete steps. (*b*) Continuous stress history broken into discrete steps.

example, the location of the maximum tensile principal stress shifted from its initial location on the inside surface to its final location on the outside surface as the stress redistribution process occurred. Such shifts are typical and complicate the analysis of the rupture process.

5.4 ANALYSIS OF THE CREEP RUPTURE OF STRUCTURES

It is also possible to analyze the creep and damage processes simultaneously. An excellent discussion of this is given in the text by Odqvist [5], from which we will take two examples. Even for the relatively simple geometries that are covered the solutions are particularly complex, especially for nonhomogeneous stress states. Creep rupture analysis is more

Analysis of the Creep Rupture of Structures

complicated than creep deformation analysis because one has to account for the spread of regions of failure from initial cracking at one point to final failure of the entire section.

5.4.1 Creep Rupture of a Tensile Specimen

Let us consider a constant load tensile test. If we ignore primary creep because the process is of long duration, then

$$\frac{d\bar{\varepsilon}}{dt} = \left(\frac{\sigma}{\sigma_c}\right)^m \qquad (5.31)$$

The natural strain and true stress are, according to Chapter 1,

$$d\bar{\varepsilon} = \frac{dl}{l} \qquad \sigma = \frac{P}{A} \qquad (5.32)$$

where l, P, and A are the current values of the length, load, and area. The strain definition becomes

$$\frac{d\bar{\varepsilon}}{dt} = \frac{1}{l}\frac{dl}{dt} \qquad (5.33)$$

Now if we recall that creep takes place at constant volume, we have

$$A_0 l_0 = A l \qquad (5.34)$$

where the zero subscript refers to the initial state. If we differentiate the constant volume relationship with respect to time, we obtain

$$A\frac{dl}{dt} + l\frac{dA}{dt} = 0$$

or

$$\frac{1}{l}\frac{dl}{dt} + \frac{1}{A}\frac{dA}{dt} = 0$$

and

$$\frac{1}{l}\frac{dl}{dt} = \frac{d\bar{\varepsilon}}{dt} = -\frac{1}{A}\frac{dA}{dt}$$

Therefore,

$$-\frac{1}{A}\frac{dA}{dt} = \left(\frac{\sigma}{\sigma_c}\right)^m = \left(\frac{P}{A\sigma_c}\right)^m \qquad (5.35)$$

Creep Rupture

Furthermore,

$$-A^{m-1} dA = \left(\frac{P}{\sigma_c}\right)^m dt$$

and integration gives

$$-\frac{1}{m}A^m = \left(\frac{P}{\sigma_c}\right)^m t + C$$

where C is an arbitrary constant. Since $A = A_0$ at $t = 0$, we find $C = -A_0^m/m$. Thus the solution becomes

$$A_0^m - A^m = \left(\frac{P}{\sigma_c}\right)^m tm \tag{5.36}$$

If we assume that the creep is ductile (see Chapter 1) the area goes to zero at a time

$$t_H^* = \frac{1}{m}\left(\frac{\sigma_c}{\sigma_0}\right)^m \tag{5.37}$$

where $\sigma_0 = P/A_0$ is the initial stress. This is known as Hoff's ductile rupture time. Note that it was derived by ignoring primary creep. On a log-log plot it would take the form

$$\log_{10} t_H^* = \text{constant} - m \log \sigma_0$$

This should be a valid representation for the left portion of a creep rupture curve where, as we pointed out in Section 1.4 the creep ruptures are ductile.

Now let us return to Eq. (5.36), but do not assume that the area goes to zero. If we use Eq. (5.37)

$$A_0^m - A^m = A_0^m \frac{t}{t_H^*}$$

or

$$1 - \left(\frac{A}{A_0}\right)^m = \frac{t}{t_H^*}$$

and

$$1-\left(\frac{\sigma_0}{\sigma}\right)^m = \frac{t}{t_H^*}$$

From this

$$\sigma = \sigma_0\left(1-\frac{t}{t_H^*}\right)^{-1/m} \quad (5.38)$$

This demonstrates that the stress goes from σ_0 to infinity as time goes from zero to the ductile rupture time, as could be expected. Suppose now that the stress can be kept constant. Then from Eq. (5.35)

$$\frac{dA}{A} = -\left(\frac{\sigma}{\sigma_c}\right)^m dt$$

and integration gives

$$\log_e A = -\left(\frac{\sigma}{\sigma_c}\right)^m t + C$$

The arbitrary constant C is found by noting that $A = A_0$ at $t=0$ so $C = \log_e A_0$. As a result

$$\log_e\left(\frac{A}{A_0}\right) = -\left(\frac{\sigma}{\sigma_0}\right)^m t$$

and

$$A = A_0 \exp\left[-\left(\frac{\sigma}{\sigma_c}\right)^m t\right] \quad (5.39)$$

Now the area varies from A_0 at $t=0$ to zero as time goes to infinity. This is different from the variation of area at constant load. To find how the load must vary we note that at constant stress

$$\frac{P_0}{A_0} = \frac{P}{A}$$

or

$$P = \frac{AP_0}{A_0}$$

and substitution into Eq. (5.39) gives

$$P = P_0 \exp\left[-\left(\frac{\sigma}{\sigma_c}\right)^m t\right] \qquad (5.40)$$

This shows how the force must vary to keep the stress constant.

Now let us extend the analysis of the uniaxial specimen to include initial elasto-plastic strains which the previous development ignored. Thus

$$\frac{d\bar{\varepsilon}}{dt} = -\frac{1}{A}\frac{dA}{dt} = \left(\frac{P}{A\sigma_c}\right)^m + \frac{d}{dt}\left(\frac{P}{A\sigma_p}\right)^{n_0} \qquad (5.41)$$

where the second term is a power law representation of the initial elasto-plastic strain caused by P. If it is elastic, we set $n_0 = 1$ and $\sigma_p = E$. Let us see how this alters the analysis and results. First, we rewrite Eq. (5.41) at constant load P as

$$-\frac{1}{A}\frac{dA}{dt} = \frac{1}{A^m}\left(\frac{P}{\sigma_c}\right)^m - n_0\left(\frac{P}{\sigma_p}\right)^{n_0} A^{-n_0-1}\frac{dA}{dt}$$

and

$$\left(-A^{m-1} + n_0\left(\frac{P}{\sigma_p}\right)^{n_0} A^{m-n_0-1}\right) dA = \left(\frac{P}{\sigma_c}\right)^m dt \qquad (5.41a)$$

This is integrated between the limits A_0 and A on the left, and from 0 to t on the right, to give

$$A_0^m - A^m + \frac{mn_0}{m-n_0}\left(\frac{P}{\sigma_p}\right)^{n_0}(A^{m-n_0} - A_0^{m-n_0}) = mt\left(\frac{P}{\sigma_c}\right)^m \qquad (5.42)$$

Eq. (5.42) is used as the starting point for two lines of inquiry. To use it for obtaining the ductile rupture time we set $t = t_p^*$ and let $A = 0$. This gives

$$t_p^* = \frac{1}{m}\left(\frac{\sigma_c}{\sigma_0}\right)^m \left(1 - \frac{mn_0}{m-n_0}\left(\frac{\sigma_0}{\sigma_p}\right)^{n_0}\right) \qquad (5.43)$$

where the initial stress $\sigma_0 = P/A_0$ has been introduced. Upon comparison of this to Eq. (5.37) we see that it is Hoff's ductile rupture time modified for elasto-plastic initial strains. Equation (5.42) can also be used to establish a brittle rupture time t_{Kp}^*, which is the time required to fail at some area $\alpha_R A_0$. Thus

$$t^*_{Kp} = \frac{1}{m}\left(\frac{\sigma_c}{\sigma_0}\right)^m\left[1-\alpha_R^m - \frac{mn_0}{m-n_0}\left(\frac{\sigma_0}{\sigma_p}\right)^{n_0}(1-\alpha_R^{m-n_0})\right] \quad (5.44)$$

The transition from ductile to brittle rupture occurs when

$$t^*_{Kp} \leq t^*_p$$

This can also be written

$$t^*_{Kp} - t^*_p = -\frac{1}{m}\left(\frac{\sigma_c}{\sigma_0}\right)^m\left[\alpha_R^m - \frac{mn_0}{m-n_0}\left(\frac{\sigma_0}{\sigma_p}\right)^{n_0}\alpha_R^{m-n_0}\right] \leq 0$$

Thus the transitional area between ductile and brittle fracture is

$$\alpha_R \geq \left(\frac{\sigma_0}{\sigma_p}\right)\left(\frac{mn_0}{m-n_0}\right)^{1/n_0} \quad (5.45)$$

In using t^*_{Kp} we do not, however, know α_R ahead of time. So we need to use the damage idea that we developed previously. We had, as in Eq. (5.20),

$$\frac{dD}{dt} = C\sigma_R^\nu$$

where $C_1\sigma^{*-n}$ has been replaced by C and n has been replaced by ν to avoid confusion with the present usage. We can also write this as

$$\frac{dD}{dt} = C\left(\frac{\sigma}{1-D}\right)^\nu$$

if we replace σ_0 in Eq. (5.22) with $\sigma = P/A$ which varies with time. Eventually this leads to

$$C\int_0^{t_F}\sigma^\nu \, dt = Ct_K\sigma_K^\nu$$

from which

$$\int_0^{t_F}\left(\frac{\sigma}{\sigma_K}\right)^\nu dt = t_K \quad (5.46)$$

Here t_F is a failure time for a variable stress σ, given t_K, the failure time for some reference value σ_K. Now since $\sigma = P/A = PA_0/A_0\,A = \sigma_0 A_0/A$

$$t_K = \int_0^{t_F} \left(\frac{\sigma_0}{\sigma_K}\frac{A_0}{A}\right)^\nu dt$$

and if we switch from time to area as the integration variable,

$$t_K = \left(\frac{\sigma_0}{\sigma_K}\right)^\nu A_0^\nu \int_{A_0}^{A_0 \alpha_R} \left(\frac{dt}{dA}\right)\frac{dA}{A^\nu}$$

If now we substitute $(dA/dt)^{-1}$ from Eq. (5.41a) into the above and integrate, we obtain

$$t_K = \left(\frac{\sigma_0}{\sigma_K}\right)^\nu \left(\frac{\sigma_c}{\sigma_0}\right)^m \left[\frac{1-\alpha_R^{m-\nu}}{m-\nu} - \left(\frac{\sigma_0}{\sigma_p}\right)^{n_0} \frac{n_0(1-\alpha_R^{m-n_0-\nu})}{m-n_0-\nu}\right] \quad (5.47)$$

where all terms are known except α_R. This can then be solved for α_R. The rupture time can then be obtained from either of Eqs. (5.43) or (5.44), depending on Eq. (5.45).

5.4.2 Extension to Multiaxial and Nonhomogeneous Stress States

The foregoing derivation involved a homogeneous uniaxial state of stress. We now outline the extension to multiaxial and nonhomogeneous states of stress.

Multiaxial Stress States. In this case the failure criterion takes the form

$$\int_0^{t_F} \left|\frac{\sigma_{\max}}{\sigma_K}\right|^\nu dt = t_K \quad (5.48)$$

This replaces Eq. (5.46) and, in fact, includes it when a uniaxial state is considered. Here $\sigma_{\max}$ has to be selected. Two common forms for multiaxial problems are to select $\sigma_{\max}$ as the greatest tensile principal stress or the greatest effective stress; see Eq. (2.22). Odqvist [5] applied this to the rupture problem of the thin pressurized tube with closed ends, and obtained ductile and brittle rupture times similar to the results we have just found for the tensile bar. We now outline the solution.

The nonvanishing stresses, effective stress, and deviators were given in Section 3.2.1. We also found in that section that $\dot\varepsilon_\theta = -\dot\varepsilon_r$, and $\dot\varepsilon_z = 0$.

Hence, the tangential strain rate can be written

$$\dot{\varepsilon}_\theta = \frac{3}{2}\left\{\left(\frac{\sigma_\theta\sqrt{3}}{2\sigma_c}\right)^{m-1}\frac{\sigma_\theta}{2\sigma_c} + \frac{d}{dt}\left[\left(\frac{\sigma_\theta\sqrt{3}}{\sigma_p}\right)^{n_0-1}\frac{\sigma_\theta}{2\sigma_p}\right]\right\} \quad (5.49)$$

This is in analogy to the elasto-plastic creep formulation of the tensile bar analysis that we have just presented. In this case, owing to the vanishing of the axial strain, the incompressibility condition can be expressed as

$$R_i h = R_{i0} h_0$$

where R_{i0} is the original inner radius, and h_0 is the original wall thickness. From this

$$\sigma_\theta = \frac{pR_{i0}}{h_0}\left(\frac{R_i}{R_{i0}}\right)^2 = \alpha^2 \sigma_0$$

where we have defined

$$\alpha = R_i/R_{i0} \qquad \sigma_0 = pR_{i0}/h_0$$

Furthermore, the tangential strain rate is related to α by

$$\frac{d\bar{\varepsilon}}{dt} = \frac{1}{R_i}\frac{dR_i}{dt} = \frac{R_{i0} d\alpha}{R_i dt} = \frac{1}{\alpha}\frac{d\alpha}{dt}$$

Thus

$$\frac{1}{\alpha}\frac{d\alpha}{dt} = \frac{\sqrt{3}}{2}\left[\left(\frac{\sigma_0\sqrt{3}}{2\sigma_p}\right)^{n_0} 2n_0 \alpha^{2n_0-1}\frac{d\alpha}{dt} + \left(\frac{\sigma_0\sqrt{3}}{2\sigma_c}\right)^m \alpha^{2m}\right] \quad (5.50)$$

Upon rearrangement and integration with $\alpha = 1$ at $t = 0$, we obtain

$$1 - \alpha^{-2m} - \frac{mn_0\sqrt{3}}{m-n_0}\left(\frac{\sigma_0\sqrt{3}}{2\sigma_p}\right)^{n_0}(1-\alpha^{-2(m-n_0)}) = m\sqrt{3}\left(\frac{\sigma_0\sqrt{3}}{2\sigma_c}\right)^m t$$

$$(5.51)$$

This corresponds to Eq. (5.36) in the tensile bar rupture problem. For a ductile failure α goes to infinity. This gives the time to failure as

$$t_p^* = \frac{1}{m\sqrt{3}} \left(\frac{\sigma_0\sqrt{3}}{2\sigma_c}\right)^{-m} \left[1 - \frac{mn_0\sqrt{3}}{m-n_0}\left(\frac{\sigma_0\sqrt{3}}{2\sigma_p}\right)^{n_0}\right] \quad (5.52)$$

Now calculate the brittle rupture time using the principal stress criterion, with $\sigma_{max} = \sigma_0 \alpha^2$. Then,

$$t_K = \int_0^{t_F} \left(\frac{\sigma_0}{\sigma_K}\right)^{\nu} \alpha^{2\nu} \, dt = \int_1^{\alpha_R} \left(\frac{\sigma_0}{\sigma_K}\right)^{\nu} \alpha^{2\nu} \frac{dt}{d\alpha} \, d\alpha$$

If we substitute $dt/d\alpha$ from Eq. (5.50) into this and integrate, we obtain

$$t_K = \frac{1}{(m-\nu)\sqrt{3}} \left(\frac{2\sigma_c}{\sqrt{3}\,\sigma_K}\right) \left(\frac{\sigma_0\sqrt{3}}{2\sigma_c}\right)^{-(m-\nu)}$$

$$\left[1 - \alpha_R^{-2(m-\nu)} - \frac{n_0(m-\nu)\sqrt{3}}{m-\nu-n_0}\left(\frac{\sigma_0\sqrt{3}}{2\sigma_p}\right)^{n_0}\left(1 - \alpha_R^{-2(m-n_0-\nu)}\right)\right]$$

$$(5.53)$$

This is used to find α_R. The time to reach α_R is then obtained from Eq. (5.51) as

$$t_{Kp}^* = \frac{1}{m\sqrt{3}} \left(\frac{\sigma_0\sqrt{3}}{2\sigma_c}\right)^{-m} \left[1 - \alpha_R^{-2m} - \frac{mn_0\sqrt{3}}{m-n_0}\left(\frac{\sigma_0\sqrt{3}}{2\sigma_p}\right)^{n_0}\left(1 - \alpha_R^{-2(m-n_0)}\right)\right]$$

$$(5.54)$$

As before, the ductile to brittle transition takes place when

$$t_{Kp}^* \leq t_p^*$$

and this occurs when

$$1 \leq \alpha_R \leq \left(\frac{m-n_0}{mn_0\sqrt{3}}\right)^{1/2n_0} \left(\frac{2\sigma_p}{\sigma_0\sqrt{3}}\right)^{1/2} \quad (5.55)$$

Note the close similarity in this development and the one for the tensile bar. This is because of the fortuitous circumstance that there is only one strain magnitude involved. If we had preferred, we could have used

$$\sigma_{max} = \sigma_e = \frac{\sigma_\theta \sqrt{3}}{2} = \frac{\sigma_0 \alpha^2 \sqrt{3}}{2}$$

throughout. This would have led to a slightly different result.

Nonhomogeneous Stress States. When the stress state is nonhomogeneous the entire section does not see the same stress, as was the case in the tensile bar and tube that we considered previously. An example of a nonhomogeneous state is the linear distribution of stress across the depth of a beam. In such a case the member will not have the same stress at every location. Failure will begin at the most highly stressed point and will propagate across the section. Here we calculate the damage according to

$$\int_0^D (1-D')^\nu \, dD' = C \int_0^t |\sigma_{max}|^\nu \, dt'$$

So that on integration

$$1 - (1-D)^{\nu+1} = \int_0^t \frac{1}{t_K} \left| \frac{\sigma_{max}}{\sigma_K} \right|^\nu dt'$$

This gives the damage incurred in a time t and a stress $\sigma_{max} = \sigma_{max}(t')$. Now there will be some point in the body where D goes to 1 at time t_I. At that point

$$t_K \sigma_K^\nu = \int_0^{t_I} |\sigma_{max}|^\nu \, dt'$$

and failure begins. For the ensuing period, failure propagates. We define

Stage of latent failure: $0 \leqslant t \leqslant t_I$

State of propagation: $t_I \leqslant t \leqslant t_{II}$

Failure of section: $t = t_{II}$

Odqvist [5] analyzes a beam in pure bending, in which the beam reaches failure first on its tensile side at $t = t_I$. He finds that the time to fail the beam to a depth h is given by

$$\frac{t}{t_I} = 1 + \frac{2}{2m-1} \left[1 - \left(\frac{h}{h_0} \right)^{2m-1} \right]$$

118   Creep Rupture

where h_0 is the original depth of the beam. For total failure, h goes to zero at $t=t_{II}$ and the formula gives

$$\frac{t_{II}}{t_I} = 1 + \frac{2}{2m-1}$$

An interesting result is obtained when the time to fail halfway through the depth, that is, $h = h_0/2$ is found. Call the time $t_{\frac{1}{2}}$. For this the formula gives

$$\frac{t_{\frac{1}{2}}}{t_I} = 1 + \frac{2}{2m-1}\left(1 - \left(\frac{1}{2}\right)^{2m-1}\right)$$

or

$$\frac{t_{\frac{1}{2}}}{t_I} \simeq 1 + \frac{2}{2m-1}$$

Since this is the same as t_{II}/t_I, we may conclude that most of the failure time is spent tearing the tensile portion of the beam.

5.5 APPROXIMATE ANALYSIS OF THE CREEP RUPTURE OF STRUCTURES WITH THE REFERENCE STRESS METHOD

In the previous section we outlined the detailed analysis of the creep rupture of a tensile bar and a thin pressurized tube. We also indicated some general rules for handling multiaxial and nonhomogeneous stress states. As can be imagined, more general cases than the ones we considered will not be so easy to solve. As a result there is a substantial motivation for the development of approximate analytical techniques.

In 1971 Penny and Marriott suggested [1] that the Reference Stress Method (RSM), as presented previously in Chapter 4, should find use in the approximate prediction of rupture times. They applied the method to a pressure vessel nozzle in a theoretical and experimental study [6]. Simultaneously, Leckie and his colleagues were deriving bounding techniques for the prediction of rupture time in terms of what they called a "representative stress." A thorough summary of their work was given by Leckie and Hayhurst [7], while a short summary was given in the previously cited study by Hayhurst et al. [8]. In this section we shall, therefore, present the application of the RSM to the problem of predicting creep rupture as accomplished by these investigators.

We begin the discussion by drawing on the excellent review given by Leckie and Hayhurst [7]. In the present context the reference stress for rupture is the stress that causes rupture of an uniaxial specimen in a time equal to the rupture time of the component being studied. As applied to creep rupture, the concept takes into account stress redistribution and multiaxial states of stress but, as before, it is strictly valid for kinematically determinate structures. The latter is not considered to be too severe a restriction since many practical structures do, in fact, satisfy the requirement of determinacy. To develop analytical procedures for the analysis of creep rupture, one has to model both the creep and the damage processes. Starting with the uniaxial state of stress, the governing equations are usually written in a form suggested by Kachanov and reported in References 1, 5, 7, and others.

$$\frac{dD}{dt} = C_1 \left(\frac{\sigma}{(1-D)\sigma^*} \right)^n = C_1 \left(\frac{\sigma}{\psi \sigma^*} \right)^n = -\frac{d\psi}{dt}$$

$$\frac{d\varepsilon^C}{dt} = \dot{\varepsilon}_0 \left(\frac{\sigma}{(1-D)\sigma^*} \right)^m = \dot{\varepsilon}_0 \left(\frac{\sigma}{\psi \sigma^*} \right)^m \quad (5.56a,b)$$

Here D is the damage, as before, and $\dot{\varepsilon}_0$, σ^*, m, and n are experimentally determined constants that we have encountered previously. We have included a new variable, known as the continuity ψ, which is equal to $1-D$. Thus, as D goes from zero to unity, the continuity goes from unity to zero.

To apply the foregoing equations to the case of constant uniaxial stress, we integrate over time from zero to t_R, the rupture time, keeping in mind that during this time ψ varies from unity to zero. This procedure gives

$$t_R = \frac{\sigma^{*n}}{C_1(n+1)\sigma^n}$$

which is a repeat of Eq. (5.24), as well as

$$\frac{\dot{\varepsilon}^C}{\dot{\varepsilon}_s t_R} = \lambda \left[1 - \left(1 - \frac{t}{t_R} \right)^{1/\lambda} \right]$$

$$\psi = \left(1 - \frac{t}{t_R} \right)^{1/(1+n)} \quad (5.57)$$

where

$$\lambda = \frac{n+1}{(n+1-m)}$$

and

$$\frac{\dot{\varepsilon}_s}{\dot{\varepsilon}_0} = \left(\frac{\sigma}{\sigma^*}\right)^m$$

is the steady creep rate of the undamaged material. It has been shown previously that these equations give excellent predictions of experimental creep rupture results.

Now let us extend the foregoing to multiaxial states of stress. To do this we first have to decide on a criterion that governs rupture in the multiaxial case. Studies of this question by Hayhurst [9], for example, indicate that the rupture behavior of common materials falls into two categories. In one category, which includes copper and certain nimonic alloys, rupture is governed by the maximum principal stress criterion. In the other category, which includes many steel and aluminum alloys, rupture is governed by a maximum shearing stress criterion. Since the reader is most likely to be interested in steel or aluminum alloys, we will restrict ourselves in what follows to shearing stress dependent behavior. Thus the multiaxial counterparts of Eqs. (5.56, 5.57) are written for such materials as [7]

$$\frac{\dot{\varepsilon}_{ij}^C}{\dot{\varepsilon}_0} = \phi^m \left(\frac{\sigma_{ij}}{\psi\sigma^*}\right) \frac{\partial \phi}{\partial(\sigma_{ij}/\psi\sigma^*)}$$

$$\frac{\dot{D}}{\sigma^*\dot{\varepsilon}_0} = \frac{\sigma_{ij}\dot{\varepsilon}_{ij}}{\sigma^*\dot{\varepsilon}_0} = \psi\phi^{m+1}\left(\frac{\sigma_{ij}}{\psi\sigma^*}\right)$$

$$-\dot{\psi} = C_1 \phi^n \left(\frac{\sigma_{ij}}{\psi\sigma^*}\right) \qquad (5.58)$$

where $\dot{D}$ is the rate of energy dissipation. We note that for this case both creep and damage are governed by functional forms ϕ. It the Tresca shearing stress criterion is chosen [10]

$$\phi = \frac{\sigma_1 - \sigma_3}{\psi\sigma^*} \qquad (5.59)$$

where $\sigma_1 > \sigma_2 > \sigma_3$, are the principal stresses. If instead we wish to use the von Mises criterion, then [10]

$$\phi = \frac{\left(\frac{3}{2} S_{ij} S_{ij}\right)^{1/2}}{\psi \sigma^*} \tag{5.60}$$

where S_{ij} are the components of the stress deviator.

Equations (5.58) are used in conjunction with the equilibrium and compatibility equations in rate form to analyze the creep rupture problem. Hayhurst [11], for example, found that in order to compute the movement of the damage front and final failure for a stretched plate with a central hole, it was necessary to use a time iterative computer solution with very small time increments. This led to heavy demands on computing power that were several orders of magnitude greater than that required for "ordinary" creep calculations. This difficulty, the associated expense, and the scatter that existed in basic creep data, let to a search for approximate techniques and resulted in the idea of a reference stress for rupture. For rupture governed by shearing stress one was able to derive the following formula for the reference stress as reported by Leckie and Hayhurst [7]

$$(\sigma_R/\sigma^*)^n = \int_V \left(\frac{\dot{D}_s}{\sigma^* \dot{\varepsilon}_0}\right)^{(m+1+n)/(m+1)} dV \bigg/ \int_V \frac{\dot{D}_s}{\sigma^* \dot{\varepsilon}_0} dV \tag{5.61}$$

where V is the volume and D_s is the stationary creep energy dissipation. In the process of deriving this formula it was also established that the rupture time that corresponds to it is a lower bound on the actual rupture time. From Eq. (5.24) this bound is obtained as

$$t_R \geq \frac{1}{C_1(1+n)} \left(\frac{\sigma^*}{\sigma_R}\right)^n \tag{5.62}$$

By using this equation, rupture times were estimated by Leckie and Martin [12] for thick pressurized cylinders in plane strain, by Hayhurst [13] for rotating discs, and by Goel [14] for thick pressurized spheres and torsion of thick cylinders. Experimental verifications of results obtained for rupture times with σ_R obtained from equations such as Eq. (5.61) were also reported by Leckie and Hayhurst [7] for torsion of cylindrical rods, pressurized cylindrical tubes, uniaxially stretched plates with central holes, and notched cylindrical tensile specimens. In all cases the theoretical results obtained from equations such as Eq. (5.61) gave rupture stresses that were conservative by 5 to 10 percent. That is to say, they predicted a lower stress to rupture a given specimen in a given time than was actually observed. The various tests reported also showed that stress redistribution is a significant factor in predicting rupture life. This was discovered when the results of using Eq. (5.61) were compared to results based in predicting rupture with either the maximum or steady state stress. Fairbairn [15] also observed, as noted in Chapter 4, that the reference stress gave the best prediction of observed rupture time in his previously cited tests on the bending of tubular beams. The significance of Eq. (5.61) can be appreciated more fully by observing that as m increases, the stationary distribution of stress approaches the plastic stress distribution. In many structures, such as plates and shells, plastic collapse occurs when at some point(s) the entire wall of the structure is at yield. In such cases it is shown that Eq. (5.61) reduces to Eq. (4.32).

Thus we have the fascinating conclusion that for materials whose rupture is governed by a shearing stress criterion, such as steel and aluminum alloys, and which have high values of the creep exponent m, the reference stress for both creep deformation and creep rupture can be approximated by the same formula.

Equation (4.32) was used as the basis of several analytical and experimental investigations. Ponter and Hayhurst [10] used it to provide analytical lower bounds on the rupture time for pure bending of beams, thick tubes under internal pressure, circular torsion bars, annular plates subject to in-plane loading, and spinning discs. Penny and Marriott [6] used Eq. (4.32) as part of a theoretical and experimental study of pressurized sphere/cylinder intersection. They tested two aluminum vessels to rupture under constant pressure and estimated the rupture times with several criteria. Their results are summarized below.

Criterion	Rupture Time (hours)	
	Model 1	Model 2
Experimental	1528	2122
Reference Stress		
Large deflection	1250	3300
Small deflection	560	1700
Stationary Stress		
Maximum effective	170	580
Maximum principal	110	375
Elastic, maximum stress	15	80

The various theoretical values in the table were obtained as follows, starting from the bottom line in the table. A linear, elastic, finite difference computer program was used to analyze each model, the maximum stress in each of the resulting calculations was identified, and the time to rupture for each value was read from a stress rupture curve for the material; next, a small deflection finite difference creep analysis was used to determine the stationary stress in each model; the maximum effective stress and the maximum principal stress were identified for each model, and the time to rupture for each was again read from the stress rupture plot; finally, two types of reference stress predictions were made by determining the limit pressure for each vessel in a separate short term collapse test. Then Eq. (4.32) was used in conjunction with the test pressure and yield stress of the vessels to calculate the reference stress, and the time to rupture for this stress was again read from the stress rupture curve of the material. This is denoted as the small deflection reference stress result. Since the deflections of the shells were actually large an attempt was also made to calculate a large deflection reference stress and corresponding rupture time. This was done at a given large deflection Δ by the following modifications of Eq. (4.32):

$$\sigma_R(\Delta) = \frac{P}{P_{\text{ult}}(\Delta)} \sigma_y \qquad (5.63)$$

where $\sigma_R(\Delta)$ is the reference stress at a large deflection Δ and $P_{\text{ult}}(\Delta)$ is the yield load at Δ. This reference stress is then used to determine the rupture time from the uniaxial stress rupture curve of the material.

It is seen that as one progresses up each column in the table the estimate of the rupture time improves. In fact, the reference stresses give the best prediction. This fits our previous statement in Chapter 4 that values based on Eq. (4.32), which takes redistribution into account, should naturally be more valid than stationary stresses in predicting rupture time. Moreover, it is seen that the elastic analysis gives predictions that are far too conservative. It is, however, unsettling to observe that the large deflection reference stress overestimates one of the experimental rupture times. Nevertheless, it is seen that the reference stress idea is useful in getting an estimate of rupture time in such a complicated but typical structure.

REFERENCES

[1] R. K. Penny, and D. Marriott, *Design for Creep*, McGraw-Hill Book Co., Ltd., London (1971).

[2] F. R. Larson, and J. Miller, A Time Temperature Relationship for Rupture and Creep Stresses, *Trans. A.S.M.E.*, vol. 174 (1952).

[3] S. S. Manson, and A. M. Haferd, A Linear Time-Temperature Relation for Extrapolation of Creep and Stress Rupture Data, N.A.C.A. TN 2890 (1953).

[4] S. S. Manson, Design Considerations for Long Life at Elevated Temperatures, A.S.M.E./A.S.T.M./I.M.E. Joint Int. Conf. on Creep, New York and London (1963).

[5] F. K. G. Odqvist, *Mathematical Theory of Creep and Creep Rupture*, Chs. 10–12 University Press, Oxford (1966).

[6] R. K. Penny, and D. L. Marriott, Creep of Pressure Vessels, *Int. Conf. on Creep and Fatigue in Elevated Temperature Applications*, Inst. Mech. Engrs. Conference Publication No. 13, Paper C204/73, Philadelphia, 1973.

[7] F. A. Leckie, and D. R. Hayhurst, Creep Rupture of Structures, *Proc. Roy. Soc. Lond. A.*, vol. 340, pp. 323–347 (1974).

[8] D. R. Hayhurst, D. A. Kelly, F. A. Leckie, C. J. Morrison, A. R. S. Ponter, and J. J. Williams, Approximate Design Methods for Creeping Structures, *Int. Conf. on Creep and Fatigue in Elevated Temperature Applications*, Inst. of Mech. Engrs., Conference Publication No. 13, Paper C219/73, Philadelphia, 1973.

[9] D. R. Hayhurst, Creep Rupture Under Multiaxial States of Stress, *J. Mech. Phys. Solids*, vol. 20, pp. 381–390 (1972).

[10] A. R. S. Ponter, and D. R. Hayhurst, Lower Bound on the Time to Initial Rupture of Creeping Structures, *J. Mech. Eng. Sci.* vol. 15, pp. 357–364 (1973).

[11] D. R. Hayhurst, Stress Redistribution and Rupture Due to Creep in a Uniformly Stretched Plate with a Central Hole, *J. Appl. Mech.*, vol. 40, pp. 244–250 (1973).

[12] J. B. Martin, and F. A. Leckie, On the Creep Rupture of Structures, *J. Mech. Phys. Solids*, vol. 20, pp. 223–238 (1972).

[13] D. R. Hayhurst, The Prediction of Creep Rupture Times of Rotating Discs Using Biaxial Damage Relationships, *J. Appl. Mech.*, vol. 40, pp. 915–920 (1973).

[14] R. P. Goel, On the Creep Rupture of a Tube and Sphere, *J. Appl. Mech.*, vol. 42, pp. 625–629 (1975).

[15] J. Fairbairn, A Reference Stress Approach to Creep Bending of Straight Tubes, *J. Mech. Eng. Sci.*, vol. 16, pp. 125–138 (1974).

EXERCISES

5.1 Derive Eq. (5.27) for fatigue.

5.2. A uniaxial specimen is subjected to the following stress history:
60,000 psi at 1500°F for 20 hours
40,000 psi at 1600°F for 100 hours
Use Fig. 5.4 to find how many additional hours at 30,000 psi and 1700° the specimen can stand.

5.3 Compare the creep strain at constant stress to the creep strain at constant load in a uniaxial test.

5.4 Repeat the analysis of the rupture of a thin pressurized tube using $\sigma_{max} = \sigma_e$.

5.5 What is the significance of setting σ_p to infinity and thus obtaining $\alpha_R \geqslant 0$ in Eq. (5.45)?

5.6 Determine an expression for the damage in a bar undergoing uniaxial stress relaxation. What is the time to rupture?

5.7 Determine an expression for the damage in a thin-walled tube undergoing torsion.

CHAPTER 6

CREEP BUCKLING

6.1 INTRODUCTION

When structures are subjected to a net compressive stress across a section they become susceptible to instability at a critical value of the compressive stress. The latter is characterized by sudden and large deflections that indicate the breakdown of internal bending resistance. This is also known as the buckling phenomenon. If the critical stress is elastic or elasto-plastic, the phenomenon is instantaneous. Such instantaneous behavior has been widely discussed in several texts on stability, notably those of Timoshenko and Gere [1], Gerard [2], and Brush and Almroth [3]. However, if the structure operates at temperatures for which creep becomes significant, the buckling may not be instantaneous; rather, it may develop at some time, called the critical time, for which the stress becomes critical. Such time dependent buckling is known as creep buckling and we discuss it in this chapter. The method of presentation is as follows: first, we solve the problem of the creep buckling of a column by analytical and empirical methods. This brings out the important features of the behavior. As with the other types of creep problems that we have discussed so far, exact analytical solutions of creep buckling are difficult to come by. Hence, in the second part we discuss approximate solutions of more typical problems that involve multiaxial stress states, such as cylindrical and spherical shells.

6.2 CREEP BUCKLING OF A COLUMN

As a demonstration of the creep buckling phenomenon we shall consider the column under an axial load. We will do this both analytically and empirically in order to set the stage for approaches to more general problems in the second part of the Chapter.

6.2.1 Analytical Solution

As a first illustration we consider a simply supported column of idealized H cross-section having two concentrated flanges each of area $A/2$ separated by a distance h. The solution was first presented by Hoff [4] and has been quoted in the texts by Boley and Weiner [5] and Odqvist [6]. It proceeds as follows: If the assembly has an initial deflection w_0 the strains in the two flanges can be expressed by

$$\varepsilon_1 = \varepsilon_0 + \left(\frac{\partial^2 w}{\partial x^2} - \frac{\partial^2 w_0}{\partial x^2} \right) \frac{h}{2}$$

$$\varepsilon_2 = \varepsilon_0 - \left(\frac{\partial^2 w}{\partial x^2} - \frac{\partial^2 w_0}{\partial x^2} \right) \frac{h}{2} \quad (6.1)$$

where ε_0 is a uniform compressive strain due to the axial load, and w is the total deflection of the assembly. ε_1 is the strain in the flange on the convex side, while ε_2 is the strain in the flange on the concave side. Subtraction of the two equations gives

$$\frac{\varepsilon_1 - \varepsilon_2}{h} = \frac{\partial^2 w}{\partial x^2} - \frac{\partial^2 w_0}{\partial x^2} \quad (6.2)$$

Moreover, by equilibrium of the stress with the external force and moment, we have

$$F = (\sigma_1 + \sigma_2)\left(\frac{A}{2}\right) = -P_0$$

$$M = (\sigma_1 - \sigma_2)\left(\frac{hA}{4}\right) = -P_0 w \quad (6.3)$$

where P_0 is the axial force and $P_0 w$ is the moment that it produces at a given section of the column. Now, since creep buckling under a constant load is a long term process, we are justified in taking the creep to be stationary and representing it by the law

$$\dot{\varepsilon}_i = \left(\frac{\sigma_i}{\sigma_c}\right)^m \qquad i=1,2 \qquad (6.4)$$

From this and Eqs. (6.3) we have

$$\frac{A\sigma_c}{2}\left(\dot{\varepsilon}_1^{1/m}+\dot{\varepsilon}_2^{1/m}\right) = -P_0$$

$$\frac{A\sigma_c}{4}\left(\dot{\varepsilon}_1^{1/m}-\dot{\varepsilon}_2^{1/m}\right) = -P_0 w \qquad (6.5)$$

where we have assumed that the initial elastic strains are negligible. If we solve Eqs. (6.5) simultaneously for $\dot{\varepsilon}_1$ and $\dot{\varepsilon}_2$ and substitute the result into the time derivative of Eq. (6.2), we obtain the governing equation of the column

$$\frac{\partial^3 (2w/h)}{\partial t\, \partial(x/L)^2} - \frac{2L^2}{h^2}\left(\frac{-P_0}{\sigma_c A}\right)^m \left[\left(1+\frac{2w}{h}\right)^m - \left(1-\frac{2w}{h}\right)^m\right] = 0 \qquad (6.6)$$

where we have used the fact that w_0 is independent of time. We will study this equation for two values of m. First we consider the linear case $m=1$ for which it reduces to

$$\frac{\partial^3 (2w/h)}{\partial t\, \partial(x/L)^2} + \frac{4L^2 P_0}{\sigma_c A h^2}\left(\frac{2w}{h}\right) = 0 \qquad (6.7)$$

This has the solution

$$w = a_0 \frac{h}{2}\left[\exp\left(\frac{4L^2 P_0}{\sigma_c A h^2 \pi^2}\right)t\right]\sin\frac{\pi x}{L} \qquad (6.8)$$

which satisfies the initial condition

$$w_0(x,0) = \frac{h}{2}a_0 \sin\frac{\pi x}{L} \qquad (6.8a)$$

Now if we examine the solution Eq. (6.8), we see that as time goes on the deflection grows without bound. Thus there is no critical condition at any finite time. Now consider the case $m=3$ with the same initial condition.

The differential Eq. (6.6) becomes

$$\frac{\partial^3(2w/h)}{\partial t \partial(x/L)^2} + \frac{4L^2 P_0^3}{h^2 \sigma_c^3 A^3}\left[3\left(\frac{2w}{h}\right)+\left(\frac{2w}{h}\right)^3\right]=0 \qquad (6.9)$$

This is a nonlinear differential equation. Hoff [4] showed that a reasonably accurate solution of it could be assumed by the one term approximation

$$\frac{2w}{h} \simeq a_1(t)\sin\frac{\pi x}{L} \qquad (6.10)$$

where $a_1(t)$ is to be determined. Substitution of the assumed form into Eq. (6.9) leads to

$$\frac{da_1}{dt} = \frac{3L^2 P_0^3}{\pi^2 h^2 \sigma_c^3 A^3}(4a_1 + a_1^3) \qquad (6.11)$$

Upon rearrangement this becomes

$$t = \frac{1}{3}\left(\frac{\pi h}{L}\right)^2\left(\frac{\sigma_c A}{P_0}\right)^3 \int_{a_0}^{a_1} \frac{da_1}{4a_1 + a_1^3} \qquad (6.11a)$$

where we have noted that the unknown a_1 goes from its initial value a_0 (the original midpoint amplitude) to the value a_1 at time t. The above equation can be integrated to give

$$t = \frac{1}{24}\left(\frac{\pi h}{L}\right)^2\left(\frac{\sigma_c A}{P_0}\right)^3 \log_e\left[\left(\frac{a_1}{a_0}\right)^2\left(\frac{4+a_0^2}{4+a_1^2}\right)\right] \qquad (6.12)$$

This can now be rearranged to yield an expression for the desired $a_1(t)$

$$\frac{a_1(t)}{a_0} = \frac{\exp(t/2k)}{\left[1+(a_0/2)^2(1-\exp t/k)\right]^{1/2}}$$

$$k = \frac{1}{24}\left(\frac{\pi h}{L}\right)^2\left(\frac{\sigma_c A}{P_0}\right)^3 \qquad (6.13)$$

From this we can see that $a_1(t)$ increases with time and that it becomes infinite at the critical time t_{CR} when the denominator goes to zero. This gives

$$t_{CR} = k \log_e \left[1 + \frac{4}{a_0^2} \right] \qquad (6.14)$$

In consideration of the solutions for $m=1$ and $m=3$, we may note the following points about the creep buckling process:

1. Whereas elastic or elasto-plastic buckling corresponds to the existence of many equilibrium configurations corresponding to loads beyond the critical, creep buckling is characterized by deflections or velocities that increase beyond all bounds.
2. Creep buckling can occur at a finite time only for a nonlinear creep law.
3. Creep buckling will occur at any axial compressive load no matter how small. The pertinent question becomes: Is t_{CR} less than or greater than the intended design life?
4. Creep buckling will occur only if the column has initial imperfections. Otherwise an infinite time for creep buckling is obtained from Eq. (6.14) with $a_0 = 0$. In practical structures there is always an initial imperfection.
5. The critical time depends strongly on the axial load but not so strongly on the initial shape.
6. Small deflection theory is not really valid near t_{CR} as the deflections are growing rapidly. We have made the assumption in order to present the behavior.
7. The Euler load for the instantaneous buckling does not appear in the above solution because the initial elastic strains were not included in the analysis. The Euler load is the instantaneous critical load of an elastic column.

6.2.2 Tangent Modulus Approach

The column buckling problem may also be solved empirically by the tangent modulus approach of Shanley [7]. In a consideration of columns that were loaded beyond the yield prior to buckling he reasoned as follows: since there are always slight imperfections in construction the progress of deflection of the column is such that both sides, the concave and the convex, are always in a state of compression. The process of loading for several increments is shown in Fig. 6.1. During strain increments 1 to 5 in the illustration, the straining is essentially uniform.

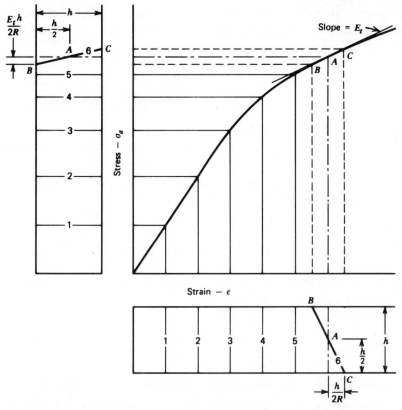

Fig. 6.1 Strain and stress distributions for an inelastic column with small initial imperfections—tangent modulus model. Courtesy of McGraw-Hill Book Co., Inc., G. Gerard, Introduction to Stability Theory (1962).

Increment 6 is approaching the critical strain, and axial straining and bending occur simultaneously. Since there is no reversal of strain on the convex side, all points lie along the path BAC on the stress-strain curve. By representing the curve locally at A by the tangent modulus the stress distribution becomes a linear function of the strain, and the critical load of the column can be obtained in the usual way, as follows, for a simply supported column with an initial shape w_0. In this case the moment equilibrium equation becomes

$$\frac{d^2w}{dx^2} + \frac{P_0}{E_t I}(w_0 + w) = 0 \tag{6.15}$$

Here w is the part of the deflection that is caused by bending, E_t is the tangent modulus, and I is the moment of inertia of the column cross-section. Now assume that the initial shape is given by the Fourier series

$$w_0 = \sum_{n=1}^{\infty} \bar{w}_n \sin \frac{n\pi x}{L} \qquad (6.16)$$

and assume that the deflection solution is given by the Fourier series

$$w = \sum_{n=1}^{\infty} w_n \sin \frac{n\pi x}{L} \qquad (6.17)$$

which satisfies the end condition of simple support. If Eqs. (6.16) and (6.17) are substituted into Eq. (6.15), we obtain

$$\sum_{n=1}^{\infty} \left[-w_n + \frac{P_0 L^2}{n^2 \pi^2 E_t I} (\bar{w}_n + w_n) \right] \sin \frac{n\pi x}{L} = 0 \qquad (6.18)$$

To obtain a nontrivial solution of this for arbitrary values of x we set the square bracket to zero and find

$$w_n = \frac{\bar{w}_n}{\dfrac{n^2 \pi^2 E_t I}{L^2 P_0} - 1} \qquad (6.19)$$

This solution becomes unbounded when the denominator goes to zero, at a load

$$P_{0_{CR}} = \frac{n^2 \pi^2 E_t I}{L^2} \qquad (6.20)$$

The lowest value of this pertains to $n=1$

$$P_{0_{CR}} = \frac{\pi^2 E_t I}{L^2} \qquad (6.20a)$$

For the elastic case $E_t = E$ and the result is called the Euler load. We have, however, considered the case where P_0/A exceeds the yield, according to Shanley's tangent modulus approach.

To apply this approach to creep we require E_t as a function of time. We have introduced a means of providing this via the isochronous stress-strain curves in Fig. 1.10. To use these in predicting creep buckling we first rearrange Eq. (6.20a)

$$E_{tCR} = \frac{P_0 L^2}{I\pi^2} = \frac{\sigma L^2 A}{I\pi^2} \qquad (6.20b)$$

where $\sigma = P_0/A$ and E_{tCR} is the critical tangent modulus. We now say that the column buckles at σ when E_t becomes E_{tCR}. The value of E_{tCR} is obtained from the load on the column and its geometry by Eq. (6.20b). We now notice from the isochronous stress-strain curves of Fig. 1.10 that at a given stress E_t decreases with time. Thus we read horizontally at the stress σ of the column and load and at some time we reach a curve with slope E_{tCR}. This time becomes the critical time for buckling. The approach, which is graphical, can be somewhat simplified if we accept the power law in steady creep, that is

$$\dot{\varepsilon} = \left(\frac{\sigma}{\sigma_c}\right)^m \qquad (6.21)$$

Then the strain is, on integration

$$\varepsilon - \varepsilon_0 = \left(\frac{\sigma}{\sigma_c}\right)^m t \qquad (6.22)$$

where ε_0 is the initial strain at $t=0$. The tangent modulus theory uses

$$E_t = \frac{d\sigma}{d\varepsilon} \qquad (6.23)$$

Thus if we differentiate Eq. (6.22) we get, for constant t,

$$d\varepsilon = \frac{m}{\sigma_c}\left(\frac{\sigma}{\sigma_c}\right)^{m-1} t\, d\sigma \qquad (6.24a)$$

from which

$$E_t = \frac{d\sigma}{d\varepsilon} = \frac{1}{\dfrac{m}{\sigma_c}\left(\dfrac{\sigma}{\sigma_c}\right)^{m-1} t} \qquad (6.24b)$$

Now Eqs. (6.20b) and (6.24b) give

$$t_{CR} = \frac{I\pi^2}{\sigma L^2 A} \frac{1}{\dfrac{m}{\sigma_c}\left(\dfrac{\sigma}{\sigma_c}\right)^{m-1}} \qquad (6.25)$$

6.3 CREEP BUCKLING OF SHELL STRUCTURES

Buckling is a phenomenon that occurs primarily in thin structures such as beams, plates, and shells where it is a definite possibility that the entire cross-section could be in compression. We already saw this for beams (columns) in the previous section. Now we take up the analysis of creep buckling in thin shell structures, particularly cylinders and spheres. The methods of solution to be presented are, of necessity, approximate.

6.3.1 The Reference Stress Method

In Chapter 4 we discussed the idea of the reference stress method (RSM) and applied it to creep deformation and stress relaxation problems. Then in Chapter 5 we showed that it could be applied to creep rupture problems. Now we will show that it can also be used in creep buckling problems.

The application of the RSM to this class of problems was first made by Penny and Marriott [8] who used it in an experimental study of hemispherical shells loaded through a central boss. They based their study on the similarity between deformations of the creeping shell and the strains in a reference stress creep test, which is the foundation of the RSM as discussed here. Thus

$$\frac{\dot{\Delta}(P,t)}{\dot{\varepsilon}(\sigma_R,t)} = \frac{\Delta_E(P)}{\varepsilon_E(\sigma_R)} \tag{6.26}$$

where

$$\Delta_E(P) = \frac{P}{S_E}$$

$$\varepsilon_E(\sigma_R) = \frac{\sigma_R}{E} \tag{6.27}$$

and ε is the total strain, P is the loading, S_E is the stiffness, $\dot{\Delta}(P,t)$ is the deformation rate of the shell, and $\dot{\varepsilon}(\sigma_R,t)$ is the strain rate in a creep test at the reference stress σ_R. When these equations are combined

$$\dot{\Delta}(P,t) = \frac{P}{\sigma_R} \frac{E}{S_E} \dot{\varepsilon}(\sigma_R,t) \tag{6.28}$$

Now, if we recall that the reference stress for a structure can be approximated by Eq. (4.32), the foregoing expression becomes

$$\dot{\Delta}(P,t) = \frac{P_L}{\sigma_y} \frac{E}{S_E} \dot{\varepsilon}(\sigma_R,t) \tag{6.29}$$

Up to this point the stiffness S_E has been the linear elastic stiffness. To make the expression apply to the large deformations that are typically involved in buckling, the elastic stiffness is replaced by the instantaneous stiffness at a given displacement Δ, as obtained from the slope of a short term load-displacement curve of the structure. Thus

$$\dot{\Delta}(P,t) = \frac{P_L}{\sigma_y} \frac{E}{S_E} \frac{S_E}{S_T} \dot{\varepsilon}(\sigma_R, t) \qquad (6.30)$$

where S_T is the tangent stiffness and this now applies to arbitrary magnitudes of displacement. As discussed in Chapter 4 the limit load P_L can be found theoretically or from a separate short term collapse test of the structure. The calculation for establishing the creep buckling time of the structure proceeds as follows:

1. Find the collapse load P_L of the structure theoretically or experimentally.
2. Calculate the reference stress with Eq. (4.32).
3. Perform a uniaxial tensile creep test at the reference stress. This gives $\dot{\varepsilon}(\sigma_R, t)$.
4. Perform a constant velocity load displacement test of the structure and obtain a curve of tangent stiffness S_T and the ratio S_T/S_E against displacement.
5. Use Eq. (6.30) to calculate the deformation rate at the current time t (starting at $t=0$).
6. Add a small time increment δt and calculate the new deflection from

$$\Delta(P, t+\delta t) = \Delta(P, t) + (\delta t)\dot{\Delta}(P, t)$$

7. Repeat steps (4) to (6) until the displacement rate $\dot{\Delta}$ begins to grow without bound. A plot of deflection against time will indicate this, and therefore gives the buckling time.

Penny and Marriott [8] conducted tests on two aluminum hemispherical shells loaded vertically along their axes through a central boss. The results for the tests, each of which involved a different load, are plotted in Fig. 6.2 along with the predictions of the RSM as described above. It is seen in the plots that in both cases the buckling time is estimated conservatively by the RSM. In one of the cases the agreement is quite close. The method, while it is conceptually quite simple, requires two items of experimental information, namely, the creep strains at the reference stress and a short term load-displacement test of the structure. It may be possible, however, to determine the latter analytically or numerically.

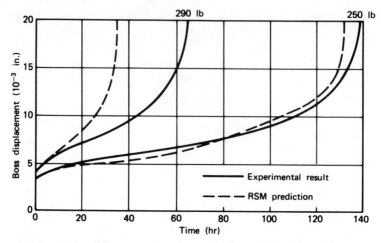

Fig. 6.2 Displacement of boss loaded hemispherical shells during creep [8].

6.3.2 The Critical Effective Creep Strain Method

The critical effective creep strain method was applied to the problem of creep buckling of shells by Chern [9], as follows:

We consider a structure that is subjected to a loading P at a constant elevated temperature T. There is an instantaneous critical buckling load P_{CR} with corresponding elastic critical stresses and strains $\bar{\sigma}_{ij}$ and $\bar{\varepsilon}_{ij}$. This critical load can be obtained by analysis or by experiment. Moreover, it can include the effects of plastic deformation and initial imperfections which reduce $\bar{\sigma}_{ij}$ and $\bar{\varepsilon}_{ij}$.

If the applied load is less than P_{CR} time varying creep strains will develop and the total strains will be

$$\varepsilon_{ij} = \varepsilon_{ij}^E + \varepsilon_{ij}^C \tag{6.31}$$

Both the elastic and the creep contributions to the total strain may vary with time because of stress redistribution. Creep buckling will then occur at a critical time when, as we showed in Section 6.2, the deformation rate tends to increase indefinitely. If, in an approximate analysis, the stress redistribution is ignored, then the stresses σ_{ij} can be determined from elastic analysis and the creep strains can be determined from the flow rule

$$\varepsilon_{ij}^C = \frac{3}{2} \frac{\bar{\varepsilon}^C}{\sigma_e} S_{ij} \tag{6.32}$$

where all symbols were defined in Chapter 2. Note that in contrast to Chapter 2 this equation is not expressed in terms of rates. It is obtained by substituting Eq. (2.25) into Eq. (2.15) and integrating with respect to time. During the integration the stress and the temperature are assumed to be constant. The effective creep strain $\bar{\varepsilon}^C$ is correlated for constant effective stress and temperature by an equation of the form

$$\bar{\varepsilon}^C = f(t, \sigma_e, T) \tag{6.33}$$

from which time can be extracted in the form

$$t = g(\bar{\varepsilon}^C, \sigma_e, T) \tag{6.34}$$

If the critical effective creep strain $\bar{\varepsilon}^C_{CR}$ at the onset of creep buckling can be estimated, the critical buckling time can also be estimated as

$$t_{CR} = g(\bar{\varepsilon}^C_{CR}, \sigma_e, T) \tag{6.35}$$

For example, if the Bailey-Norton Law is used,

$$\bar{\varepsilon}^C = A(T)\sigma_e^m t^n \tag{6.36}$$

then

$$t_{CR} = \left(\frac{\bar{\varepsilon}^C_{CR}}{A\sigma_e^m}\right)^{1/n} \tag{6.37}$$

To estimate the critical creep strain for use in equations such as Eqs. (6.34) or (6.37) we have several methods available. These will now be presented.

Secant Modulus Method. The elastic stress-strain law from Chapter 2 is, by rearrangement, expressible in the form

$$\varepsilon_{ij}^E = \frac{3}{2E} S_{ij} - \frac{1-2\nu}{2E}(\sigma_{ij} - \sigma_{kk}\delta_{ij}) \tag{6.38}$$

Now if we substitute this and the creep strains from Eq. (6.32) into Eq. (6.31) we obtain

$$\varepsilon_{ij} = \frac{3}{2}\left(\frac{1}{E} + \frac{\bar{\varepsilon}^C_{CR}}{\sigma_e}\right) S_{ij} - \frac{1-2\nu}{2E}(\sigma_{ij} - \sigma_{kk}\delta_{ij}) \tag{6.39}$$

where the critical buckling strains have been used. It is assumed that a secant modulus and associated Poisson's ratio ν_s can be defined so that the *form* of the elastic strain Eq. (6.38) also holds for the *total* strains, or

$$\varepsilon_{ij} = \frac{3S_{ij}}{2E_s} - \frac{1-2\nu_s}{2E_s}(\sigma_{ij} - \sigma_{kk}\delta_{ij}) \tag{6.40}$$

Then by equating Eqs. (6.39) and (6.40) the result is

$$\frac{3}{2}\left(\frac{1}{E} + \frac{\bar{\varepsilon}_{CR}^C}{\sigma_e} - \frac{1}{E_s}\right)S_{ij} = \left(\frac{1-2\nu}{2E} - \frac{1-2\nu_s}{2E_s}\right)(\sigma_{ij} - \sigma_{kk}\delta_{ij}) \tag{6.41}$$

For arbitrary states of stress this equation can be satisfied by setting the quantities in the parentheses equal to zero. This gives

$$\bar{\varepsilon}_{CR}^C = \frac{\sigma_e}{E}\left[\frac{E}{E_s} - 1\right] \tag{6.42a}$$

$$\nu_s = \frac{1}{2} - \left(\frac{1}{2} - \nu\right)\frac{E_s}{E} \tag{6.42b}$$

This was also derived for elasto-plastic solids by Nadai [10] and Stowell and Richard [11].

In addition to Eqs. (6.42a,b) one more equation is needed to determine $\bar{\varepsilon}_{CR}^C$, E_s, and ν_s for a given σ_e, E, and ν. To derive this we first note that the critical buckling load P_{CR} can be expressed as

$$P_{CR} = g_1(E, \nu)g_2(h, R, d, \omega) \tag{6.43}$$

where h, R, and d are geometric parameters, and ω is an imperfection parameter. The equation embodies the assumption that the dependence on the material properties is separable from the dependence on the geometric properties. Thus if we now speak of an equivalent structure with material properties E_s and ν_s but the same dimensions, then we can obtain the ratio

$$\frac{g_1(E_s, \nu_s)}{g_1(E, \nu)} = q \tag{6.44a}$$

and

$$q = \frac{P}{P_{CR}} \tag{6.44b}$$

where q is a dimensionless parameter that describes the applied load.

When E_s and ν_s are found from Eqs. (6.44) and (6.42b), $\bar{\varepsilon}_{CR}^C$ follows from Eq. (6.42a). In a series of numerical problems concerning cylindrical and spherical shells Chern [9] noted that Eq. (6.44a) could be approximated by

$$\frac{E_s}{E} = q \tag{6.44c}$$

from which Eq. (6.42a) becomes

$$\bar{\varepsilon}_{CR}^C = \frac{\sigma_e}{E}\left(\frac{1}{q} - 1\right) \tag{6.44d}$$

At this point we observe that the critical strains ε_{ij} determined by this method are different from the instantaneous buckling values $\bar{\varepsilon}_{ij}$.

Constant Critical Strain Method. In this method it is assumed that the critical buckling strains corresponding to the instantaneous buckling and the creep buckling are the same. Thus we equate the creep strains from Eq. (6.32) and the elastic strains from Eq. (6.38)

$$\frac{3}{2}\frac{\bar{\varepsilon}_{CR}^C}{\sigma_e}S_{ij} = \frac{1+\nu}{E}(\bar{S}_{ij} - S_{ij}) + \frac{1-2\nu}{3E}(\bar{\sigma}_{kk} - \sigma_{kk})\delta_{ij} \tag{6.45}$$

where $\bar{S}_{ij}$ and $\bar{\sigma}_{kk}$ pertain to the critical stress of instantaneous buckling. When the latter are substituted into Eq. (6.45) the critical strain $\bar{\varepsilon}_{CR}^C$ can be determined. The differences in Eq. (6.45) are there to indicate that the creep strain develops in going from the current stress σ_{ij} to the critical stress $\bar{\sigma}_{ij}$. Eq. (6.45), as it stands, can be used to evaluate creep buckling due to an individual dominant stress component. Alternately, if we proceed on the basis of the equality of the strain deviators, Eq. (6.45) is replaced by

$$\frac{3}{2}\frac{\bar{\varepsilon}_{CR}^C}{\sigma_e}S_{ij} = \frac{1+\nu}{E}(\bar{S}_{ij} - S_{ij}) \tag{6.46a}$$

This, on rearrangement, becomes

$$\left(\frac{3}{2}\frac{\varepsilon_{CR}^C}{\sigma_e} + \frac{1+\nu}{E}\right)S_{ij} = \frac{1+\nu}{E}\bar{S}_{ij} \tag{6.46b}$$

Now if we square both sides and recall the definition of σ_e, we get the result

$$\bar{\varepsilon}_{CR}^C = \frac{\sigma_e}{E} \frac{2(1+\nu)}{3} \left(\frac{\bar{\sigma}_e}{\sigma_e} - 1 \right) \qquad (6.47)$$

where $\bar{\sigma}_e$ is the effective stress that is associated with the critical stresses $\bar{\sigma}_{ij}$.

The significance of Eq. (6.47) is that it represents the assumption that both the instantaneous buckling and the creep buckling have the same critical strain, which is defined here in terms of the deviatoric strains. This is so because the two sides of Eq. (6.46b) represent the critical deviatoric strains at the onset of buckling in each case.

Isochronous Stress Strain Method. This method treats the creep buckling problem as an equivalent plastic buckling problem in which the nonlinear instantaneous stress-strain relation is replaced with the isochronous stress-strain curve corresponding to the critical buckling time. We have, in fact, already used this in Section 6.2.2, where it was applied to the buckling of a column. The method is attractive because many solutions for elasto-plastic buckling are available [12]. It has been applied to beam columns [13], [14] and cylindrical shells [15].

In terms of what has already been presented here the secant modulus method can be extended to the isochronous stress-strain method. By definition of the method, the critical effective stress and strain at the onset of creep buckling should be on the particular isochronous stress-strain curve corresponding to the critical buckling time t_{CR}. The critical effective stress σ_e is considered as the critical stress of instantaneous buckling for the equivalent structure and can be expressed by

$$\sigma_e = g_3(E_t, E_s, \nu_s) \, g_4(h, R, d, \omega) \qquad (6.48)$$

Again this expresses separability of geometric and material effects. It is similar to Eq. (6.43) except that the equivalent parameters E_s, E_t, and ν_s are used to account for the plastic deformation. ν_s is defined by Eq. (6.42b), whereas

$$E_s = \frac{\sigma_e}{\bar{\varepsilon}} \qquad E_t = \frac{d\sigma_e}{d\bar{\varepsilon}} \qquad (6.49)$$

which have to be satisfied by the isochronous stress-strain curve for creep buckling. Since the total effective strain is given by

$$\bar{\varepsilon} = \bar{\varepsilon}^E + \bar{\varepsilon}^C = \frac{\sigma_e}{E} + \bar{\varepsilon}^C \qquad (6.50)$$

Eqs. (6.49) give

$$\bar{\varepsilon}_{CR}^{C} = \frac{\sigma_e}{E}\left(\frac{E}{E_s} - 1\right) \quad (6.51a)$$

$$\frac{1}{E_t} = \frac{1}{E} + \frac{d\bar{\varepsilon}^C}{d\sigma_e} \quad (6.51b)$$

at the critical creep strain for buckling. Eq. (6.51a) is identical to Eq. (6.42a) from the secant modulus method. To determine E_t from Eq. (6.51b) we first need the second term. This is obtained from Eq. (6.51a) with time eliminated, as in Eq. (6.34), and results in an expression of the form

$$\frac{d\bar{\varepsilon}^C}{d\sigma_e} = k(\sigma_e, \bar{\varepsilon}_{CR}^C) \quad (6.52)$$

Now on substitution of E_t from Eq. (6.51b) and ν_s from Eq. (6.42b) into Eq. (6.48), E_s can be determined. Then $\bar{\varepsilon}_{CR}^C$ is obtained from Eq. (6.51a). Generally, this procedure requires iteration because Eq. (6.52) may not be explicit in $\bar{\varepsilon}_{CR}^C$. The procedure simplifies if the power law

$$\bar{\varepsilon}^C = A(T)\,\sigma_e^m B(t) \quad (6.53)$$

is used, where $B(t)$ is a function of time. With this, Eq. (6.52) becomes

$$\frac{d\bar{\varepsilon}^C}{d\sigma_e} = \frac{m\bar{\varepsilon}_{CR}^C}{\sigma_e} \quad (6.54a)$$

and on substitution of (6.51a), it becomes

$$\frac{d\bar{\varepsilon}^C}{d\sigma_e} = m\left(\frac{1}{E_s} - \frac{1}{E}\right) \quad (6.54b)$$

Furthermore, Eq. (6.51b) becomes

$$\frac{1}{E_t} = \frac{m}{E_s} - \frac{m-1}{E} \quad (6.54c)$$

Now, no iteration is required.

6.3.3 Comparison of Methods

Chern [9] compared the above methods for several structures: the cylindrical shell under external pressure, the cylindrical shell under axial load, the spherical shell under external pressure, and the boss loaded hemispherical shell. In each case the predictions were compared to available experimental data. For brevity we will discuss the results for the axially compressed cylinder because it is the only case where all three methods were compared. We will also discuss the boss loaded, hemispherical shell because it affords us a comparison to the results of the reference stress method that was applied to this problem in Section 6.3.1.

The Axially Compressed Cylindrical Shell. For a circular cylindrical shell of mean radius R and thickness h that buckles in an axisymmetric mode under an axial compressive load P per unit cross-sectional area, the buckling load is given by

$$P_{CR} = \frac{\alpha E h}{R\sqrt{3(1-\nu^2)}} \tag{6.55}$$

where α is a modification factor that assumes the value unity for a thin, perfect, elastic cylinder. Experimental values give $0.1 \leqslant \alpha \leqslant 0.6$. Chern [9] assumed $\alpha = 0.6$ for the analysis. The only nonzero stress component in the shell is the axial stress

$$\sigma_x = -P \tag{6.56a}$$

Therefore,

$$\sigma_e = P \tag{6.56b}$$

According to the secant modulus method, we obtain from Eqs. (6.44a) and (6.55) for this problem

$$q = \left(\frac{1-\nu^2}{1-\nu_s^2}\right)^{1/2} \frac{E_s}{E} \tag{6.57}$$

This, together with Eq. (6.42b) for ν_s can be used to determine E_s as

$$\frac{E}{E_s} = -\frac{2}{3}\left(\frac{1}{2}-\nu\right) + \left[\frac{16}{9}\left(\frac{1}{2}-\nu\right)^2 + \frac{4}{3}\frac{1-\nu^2}{q^2}\right]^{1/2} \tag{6.58}$$

With this value of E_s and Eq. (6.56b), ν_s and $\bar{\varepsilon}_{CR}^C$ can be calculated from Eqs. (6.42a, b). The ratio between ε_x and $\varepsilon_{x,CR}$ can be found as

$$\frac{\varepsilon_x}{\varepsilon_{x,CR}} = \left(\frac{1-\nu^2}{1-\nu_s^2}\right)^{1/2} \tag{6.59}$$

For the constant critical strain method using Eq. (6.45) and the axial strain, Chern [9] finds that

$$\bar{\varepsilon}_{CR}^C = \frac{\sigma_e}{E}\left(\frac{1}{q}-1\right) \tag{6.60}$$

whereas, using the effective strain, the result is given by Eq. (6.47).

Finally, for the isochronous stress-strain method, Eq. (6.48) takes the form [12]

$$\sigma_e = \left(\frac{E_t E_s}{3(1-\nu_s^2)}\right)^{1/2} \frac{h}{R} \tag{6.61}$$

Then, if the power law Eq. (6.53) is adopted, E_t can be eliminated from Eq. (6.61) with Eq. (6.54c) to give

$$x = \frac{3}{2} m^{1/2} y [1 + z(x)]^{1/2} \tag{6.62}$$

where

$$x = \frac{E_s}{E} \qquad y = \frac{PR}{Eh} \tag{6.62a}$$

and

$$z(x) = \frac{4x}{3m}\left\{(m-1)\left(\frac{1}{2}-\nu\right)^2 x^2 - \left[m\left(\frac{1}{2}-\nu\right)^2 + (m-1)\left(\frac{1}{2}-\nu\right)\right]x\right.$$
$$\left. -\left(\frac{m}{4}-\frac{3}{4}+m\nu\right)\right\} \tag{6.62b}$$

The system in Eq. (6.62) has to be solved for x iteratively, starting with the trial value

$$x_0 = \frac{3}{2} m^{1/2} y \tag{6.62c}$$

which is obtained by letting $z = 0$. Then a value for $z(x_0)$ is calculated and a new value x_1 is calculated from Eq. (6.62). This is continued until successive values x_n and x_{n+1} change less than some small, preset amount. Now we can find E_s from Eq. (6.62a), and $\bar{\varepsilon}_{CR}^C$ follows from Eq. (6.51a).

The three approaches that have just been developed were applied to some experiments that were performed by Samuelson [17] on aluminum

TABLE 6.1 Comparison of Buckling Time Predictions (Hours) For Aluminum Cylinders Under Axial Compression [9].

Nominal Axial Stress, kg/mm	Radius to Thickness Ratio	Samuelson [17]		Present Predictions			
		Theory	Test	Secant Modulus	Critical Axial Strain	Critical Effective Strain	Isochronous Stress-Strain Curve
8	32	90.	73.	139.	127.	110.	98.6
9.53	143	1.5	1.8	2.84	2.64	2.29	3.86
9.40	144	1.5	3.0	3.13	2.91	2.52	4.18
10	32	25.5	24.5	36.6	33.4	28.9	26.1
10	32	25.5	25.5	36.6	33.4	28.9	26.1
10.2	116	2.6	3.6	3.49	3.23	2.80	3.68
10	119	2.6	4.8	3.77	3.49	3.02	4.01
10	116	2.6	5.9	4.08	3.77	3.27	4.20
12	33	7.7	8.3	11.6	10.6	9.22	8.40
12	33	7.7	8.0	11.6	10.6	9.22	8.40
12	33	7.7	8.2	11.6	10.6	9.22	8.40
12.1	32	7.7	7.2	11.5	10.5	9.12	8.30
12	58	3.2	3.6	5.12	4.71	4.08	3.99
12	59	3.2	3.9	4.98	4.57	3.96	3.89
12	59	3.2	3.7	4.98	4.57	3.96	3.89
12	77	1.7	1.8	3.01	2.77	2.40	2.59
12	79	1.7	2.1	2.85	2.62	2.27	2.49
12	78	1.7	2.4	2.93	2.70	2.34	2.54
12	77	1.7	3.4	3.01	2.77	2.40	2.59
12	82	1.7	2.2	2.62	2.41	2.09	2.34
12	84	1.7	2.2	2.47	2.28	1.98	2.24
12	111	0.4	0.6	1.04	0.97	0.84	1.34
12	115	0.4	0.73	0.89	0.82	0.71	1.24
12	110	0.4	0.94	1.08	1.00	0.87	1.36
12	114	0.4	0.75	0.92	0.86	0.74	1.27
12	118	0.4	0.92	0.78	0.72	0.63	1.18
Mean Ratio			0.763	1.28	1.18	1.02	1.21

cylinders. In these tests $A = 44(10^{-10})$, $m = 5.8$, and $n = 1$ for use with Eq. (6.36), and $\nu = 0.3$, $E = 4.8(10^3)$ kg/mm². The results of the tests and the present predictions are given in Table 6.1 along with a theoretical prediction by Samuelson [17]. Chern [9] defined a mean ratio to evaluate the prediction as follows:

$$MR = \frac{1}{n} \sum_{i=1}^{n} \frac{f_i}{F_i} \qquad (6.63)$$

Here n is the number of load cases considered, f_i is the predicted creep buckling time for the i-th case, and F_i is the measured creep buckling time for the i-th case. A perfect prediction would have $MR = 1$. In Table 6.1 it is thus seen that Samuelson's [17] own predictions underestimate the buckling time by 24% while the secant modulus, critical axial strain, critical effective strain, and isochronous stress-strain methods overestimate the measurements by 28%, 18%, 2%, and 21%, respectively.

The Boss Loaded Hemispherical Shell. In Section 6.3.1 we gave the results of experiments and reference stress predictions as obtained for this problem by Penny and Marriott [8]. Chern [9] analyzed these tests with the secant modulus and the constant critical strain methods. We now present these.

The instantaneous critical buckling load for these shells as obtained by Baur et al. [18] is

$$P_{CR} = 15.57 \, E\left(\frac{h}{R}\right) d^2 \left(\frac{h^2}{12R^2}\right)^{0.6368} \qquad (6.64)$$

where h, and R are the thickness and radius of the shell, and d is the diameter of the rigid boss. Under an applied load $P < P_{CR}$, the elastic membrane stresses at the critical location (the root of the boss) are

$$\sigma_\phi = -\sigma_\theta = \frac{-2PR}{\pi h d^2} \qquad (6.65a)$$

for which the effective stress is

$$\sigma_e = \frac{2\sqrt{3} \, PR}{\pi h d^2} \qquad (6.65b)$$

For the secant modulus method, the critical effective strain is found to be

$$\bar{\varepsilon}_{CR}^C = \frac{\sigma_e}{E}\left(\frac{1}{q} - 1\right) \qquad (6.66)$$

since Eq. (6.64) gives us, according to Eq. (6.44a),

$$\frac{E_s}{E} = q$$

The ratio between the critical strain for creep buckling ε_ϕ (equals $-\varepsilon_\theta$) and that for instantaneous buckling $\bar{\varepsilon}_\phi$ (equals $-\bar{\varepsilon}_\theta$) becomes

$$\frac{\varepsilon_\phi}{\bar{\varepsilon}_\phi} = \frac{\frac{3}{2} - \left(\frac{1}{2} - \nu\right)q}{1+\nu} \tag{6.67}$$

The corresponding result for constant critical strain is given by Eq. (6.47). For the tests by Penny and Marriott [8] the dimensions were $R = 3$ in., $h = 0.046$ in., and $d = 0.75$ in. The creep data for use in Eq. (6.36) were $A = 1.44(10^{-12})$, $m = 2$, and $n = 0.4$. In addition, $E = 0.5(10^6)$ psi, and $\nu = 0.33$.

The predictions of creep buckling time are compared with the results for the two tests and the reference stress predictions [8] that were described in Section 6.3.1 in Fig. 6.3. The values for the mean ratio defined by Eq. (6.63) are also given. There it is seen that the RSM underpredicts the two experimental results by 26%, and the critical strain and secant modulus methods overpredict the results by 41% and 90%, respectively.

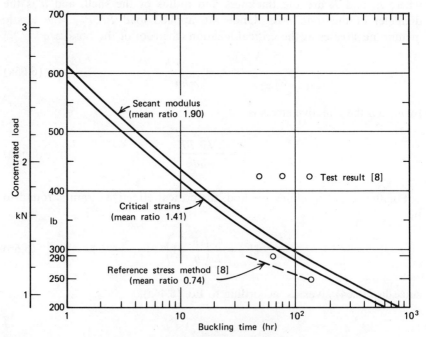

Fig. 6.3 Buckling time for an aluminum hemisphere loaded through a rigid boss at the apex [9]. Courtesy of the American Society of Mechanical Engineers.

Conclusions. On the basis of his studies Chern [9] concluded that the constant critical strain method is the simplest to use and has an overall accuracy that is slightly better than the other critical effective creep strain methods that have been described. The RSM gave the best prediction in the only available study in which it was used. It is not very simple to use, as we stated in Section 6.3.1, because of the experimental data that it requires for its implementation. However, unlike the other methods, it was conservative in the one case where it was applied. That is, it predicted buckling earlier than it actually occurred. In Chern's [9] comparison of four series of experiments the effective creep strain methods did not always come out on the same side of the measured results.

It should be emphasized that the three critical effective strain methods, as derived here, pertain to situations where the stress, load, and temperature are constant. Chern [9] suggests that if variable loads, for example, are to be treated, the concept of cumulative damage considered in Section 5.3.2 could apply here as well. For buckling the criterion for multiple load levels would be

$$\sum_{i=1}^{n} \frac{t_i}{t_{CRi}} = 1 \qquad (6.68)$$

where t is the time at the i-th load level and t_{CRi} is the critical time to creep buckling at the i-th load level. Nevertheless, it is still assumed that there is no stress redistribution at each load level. This is a limitation of the critical effective strain approach. Variable loads can be accommodated in the RSM, as discussed in Chapter 4, by carrying out the reference stress test for a variable stress history.

6.4 CLOSURE

Creep buckling is a complicated non-linear problem. Given enough time, any net compressive stress across a section could lead to buckling if the structure operates at a temperature level for which creep is significant. In this chapter we have presented a few currently available approximate approaches to the problem. As with all the other problems we have mentioned so far, a numerical approach is also useful in creep buckling. In Chapter 8 we discuss the application of digital computers to the creep buckling of structures. One situation for which a numerical solution gives the only hope for obtaining an answer is when one wishes to consider stress redistribution during the creep buckling process.

REFERENCES

[1] S. P. Timoshenko, and J. M. Gere, *Theory of Elastic Stability*, 2nd ed., McGraw-Hill Book Co., Inc., New York, (1961).

[2] G. Gerard, *Introduction to Structural Stability Theory*, McGraw-Hill Book Co., Inc., New York (1962).

[3] D. O. Brush, and B. O. Almroth, *Buckling of Bars, Plates and Shells*, McGraw-Hill Book Co., Inc., New York (1975).

[4] N. J. Hoff, Buckling and Stability, 41st Wilbur Wright Memorial Lecture, *J. Roy. Aero. Soc.*, vol. 58, pp. 1–52 (1954).

[5] B. A. Boley, and J. H. Weiner, *Theory of Thermal Stresses*, John Wiley & Sons, Inc., New York (1960).

[6] F. K. G. Odqvist, *Mathematical Theory of Creep and Creep Rupture*, 2nd ed., Clarendon Press, Oxford, (1974).

[7] F. R. Shanley, Inelastic Column Theory, *J. Aeronaut. Sci.*, vol. 14, pp. 261–267 (1947).

[8] R. K. Penny and D. L. Marriott, Creep Buckling of Boss Loaded Spherical Shells, *Proc. First International Conference on Pressure Vessel Technology*, Delft, vol. II (1969), pp. 861–867.

[9] J. M. Chern, A Simplified Approach to the Predictions of Creep Buckling Time in Structures, in R. S. Barsoum, editor, *Simplified Methods of Pressure Vessel Analysis*, A. S. M. E., New York, pp. 99–127 (1978).

[10] A. Nadai, *Theory of Flow and Fracture of Solids*, McGraw-Hill Book Co., Inc., New York, 2nd ed. (1950).

[11] E. Z. Stowell, and A. P. Richard, Poisson's Ratio and Volume Changes for Plastically Orthotropic Material, NACATN 3736 (1956).

[12] G. Gerard, and H. Becker, *Handbook of Structural Stability, Part III Buckling of Curved Plates and Shells*, NACATN 3783 (1957).

[13] F. R. Shanley, *Weight-Strength Analysis of Aircraft Structures*, Dover, New York, pp. 263–385 (1960).

[14] R. L. Carlson, Time Dependent Tangent Modulus Applied to Column Creep Buckling, *J. Appl. Mech.*, vol. 78, pp. 390–394 (1956).

[15] J. C. Gerdeen, and V. K. Sazawal, A Review of Creep Instability in High Temperature Piping and Pressure Vessels, Welding Research Council Bulletin No. 195, pp. 33–56 (1974).

[16] L. H. Donnell, and C. C. Wan, Effect of Imperfections on Buckling of Thin Cylinders and Columns Under Axial Compression, *J. Appl. Mech.*, vol. 17, p. 73 (1950).

[17] L. A. Samuelson, Experimental Investigation of Creep Buckling of Circular Cylindrical Shells Under Axial Compression and Bending, *J. Eng. Industry*, A. S. M. E., vol. 90, pp. 589–595 (1968).

[18] L. Baur, E. L. Reiss, and H. B. Keller, Axisymmetric Buckling of Rigidly Clamped Hemispherical Shells, *Int. J. Non-Linear Mech.*, vol. 8, pp. 31–39 (1973).

EXERCISES

6.1 Set $m=3$ in Eq. (6.25) and compare the result to that of Eq. (6.14). Discuss the differences in results.

6.2 Include elastic strains in the analysis of Section 6.2.1 and attempt to carry out the solution.

6.3 Verify that substitution of Eq. (6.10) into (6.9) leads to Eq. (6.11).

6.4 Derive Eqs. (6.38) and (6.42a,b).

6.5 Derive Eqs. (6.45), (6.46a) and (6.47).

6.6 For a long perfectly circular cylindrical shell of mean radius R and thickness h the critical external pressure at the onset of elastic buckling is

$$P_{CR} = \frac{E\left(\dfrac{h}{R}\right)^3}{4(1-\nu^2)}$$

Determine the critical time to creep buckling by using the secant modulus and constant critical strain methods.

6.7 For a spherical shell of thickness h and mean radius R the critical external pressure at the onset of elastic buckling is

$$P_{CR} = \frac{2\alpha E\left(\dfrac{h}{R}\right)^2}{[3(1-\nu^2)]^{1/2}}$$

where α is a known reduction factor. Calculate the time to creep buckling with the two constant critical strain approaches.

6.8 Derive Eqs. (6.58), (6.59), and (6.60).

CHAPTER 7

CREEP RATCHETTING

7.1 INTRODUCTION

In many industrial structures it happens that there is a sustained mechanical load on which a cyclic thermal load is superimposed. Examples of this are pressure vessels and turbine discs. In the pressure vessel the sustained mechanical load is pressure in the vessel, while the cyclic thermal load is caused by fluctuations in the temperature of the fluid caused by, for example, changes in the power level of a nuclear reactor. In the turbine disc and other components of gas turbines, the mechanical load is caused by the centrifugal force, while the thermal cycles are again caused by fluctuation in the power level.

In this chapter we show that such situations lead to a variety of responses in the structures that are involved, depending on the magnitudes of both the sustained mechanical stress and the accompanying fluctuating thermal stress. The simplest response is the elastic one. However, if the stresses exceed yield, several possibilities arise. First, the structure may shake down to elastic action, that is, after a one-half cycle excursion into the plastic regime the behavior for the remainder of the thermal cycles is elastic. Second, the structure may incur alternating plasticity. In this behavior plastic strains are applied and removed, and failure eventually occurs by low cycle fatigue. Finally, the third possibility is that ratchetting could occur. This results in an increment of plastic strain during each cycle and eventually leads to unacceptable changes in the dimensions of the parts. These could be growth or shrinkage, depending on the arrangement. The ratchetting behavior is of interest to us in this chapter, although the other modes will be discussed as well. It was first analyzed for nuclear reactor pressure vessels by Miller [1] and later by Edmunds and Beer [2], Burgreen [3], [4] and Bree [5]. Gatewood analyzed ratchetting in aircraft structures [6]. The behavior that has just been outlined can occur at room temperature. However, as soon as the temperature level becomes high enough creep and relaxation enter the process as well. In fact it has been shown by Bree [5], [7] that, at elevated temperature, only two types of response to the loads we have mentioned are possible. That is, either the behavior is elastic or there is some type of ratchetting. Thus at elevated temperature there is no shakedown or alternating plasticity for this type of loading.

We should add here that mechanical ratchetting can also occur as an interaction between sustained tensile loads and alternating bending loads. It can also occur because of thermal cycling without mechanical load due to the difference in yield strength at the hot and cold extremes of the cycle. In this chapter we will not discuss these types of ratchetting. We will focus on the thermal ratchetting, with and without creep, that has been outlined. As part of this we shall also bring in alternating plasticity and shakedown on the way to obtaining the ratchetting behavior. This requires that plasticity be considered, and we shall bring in whatever is necessary from that theory to make the treatment as complete as possible.

Since the analysis of arbitrary structures under arbitrary steady mechanical stresses and fluctuating thermal stresses is extremely complicated and generally requires a computer solution, we use as an illustration of the problem the idealization proposed by Bree [5], [7] for a thin cylindrical tube. This is felt to be a valid representation for many practical problems, and it brings out all the features that we wish to emphasize.

7.2 A THIN TUBE SUBJECTED TO CONSTANT INTERNAL PRESSURE AND CYCLIC THERMAL STRESSES

In this section we discuss the so-called Bree problem [5], [7]. This is done in two phases. First, we ignore creep and relaxation and give an elastoplastic analysis. Then we consider the effects of creep and relaxation. The problem to be considered is that of a cylindrical tube of mean radius R and wall thickness h that is closed at the ends. The tube is subjected to an internal pressure p and a temperature drop across its wall that is cycled between ΔT and zero. In such a tube the average hoop and axial stresses due to the pressure are, from equilibrium

$$\sigma_\theta = \sigma_p \qquad \sigma_z = \frac{\sigma_p}{2} \tag{7.1}$$

where

$$\sigma_p = \frac{pR}{h} \tag{7.2}$$

These stresses are large in comparison to the radial stress, which is, therefore neglected. The temperature distribution across the wall is assumed to be linear, that is,

$$T = \frac{-\Delta T x}{h} \tag{7.3}$$

during the first half of each cycle (the so-called start-up) and zero during the second half of each cycle (the so-called shut down), where x is measured radially outward from the midwall of the tube. Thus, in this case, the inner surface is hotter than the outer surface of the tube. For simplicity an element of the tube wall can be considered as a flat slab subjected to stresses σ_p and $\sigma_p/2$ in the two directions and *prevented from bending* in the two directions during the cycling of the temperature. The elastic solutions for the stresses in the tube can be found to be

$$\sigma_\theta = \sigma_p + \frac{2x\sigma_t}{h}$$

$$\sigma_z = \frac{\sigma_p}{2} + \frac{2x\sigma_t}{h} \tag{7.4a}$$

where

$$\sigma_t = \frac{E\alpha\Delta T}{2(1-\nu)} \tag{7.4b}$$

is the maximum value of the elastic thermal stress, which varies between $-\sigma_t$ at $x=-h/2$ and σ_t at $x=h/2$. If the tube is allowed to yield, the solution for this simple situation can no longer be obtained in closed form owing to the biaxiality. Hence, Bree [5] made the additional assumption that, since the hoop stress is the greater of the two stresses acting, the axial stress would be zero. Thus the problem is reduced to that of a slab with a mean stress σ_p acting in one direction only, corresponding to the hoop direction, and prevented from bending in that direction during application and removal of the temperature. Now the uniaxial stress is given by

$$\sigma = \sigma_p + \frac{E\alpha\Delta T_1 x}{h} \quad (7.5)$$

where the suffix θ has been dropped because the stress state is one-dimensional. ΔT_1 denotes the temperature difference in the uniaxial model. On comparing this with the first of Eqs. (7.4a) we find that now there is no factor $(1-\nu)$ present in the thermal stress. However, Bree [5] argued that the stress in the uniaxial model could be made to simulate the hoop stress in the tube with reasonable accuracy by taking

$$\Delta T_1 = \frac{\Delta T}{1-\nu} \quad (7.5a)$$

On substituting for ΔT_1 in Eq. (7.5), the stress takes the form

$$\sigma = \sigma_p + \frac{2x\sigma_t}{h} \quad (7.5b)$$

which, for elastic behavior, equals the hoop stress in the tube. We should emphasize that the uniaxial assumption is made to permit a relatively simple solution. It is not claimed that the neglect of the axial stress is exact. Under the hoop stress, the tube grows radially while if the axial stress were retained, its effect would be to make the tube contract radially. Thus the neglect of axial stress should yield results that are conservative, that is, greater than would be obtained by an actual solution.

7.2.1 The Elasto-Plastic Problem

Initially, we assume that no creep or relaxation occurs. The problem is thus elasto-plastic and is set up as follows [5]. If the total strain in the direction of the stress σ is denoted by ε, the condition that *there be no bending* is expressed by

$$\varepsilon = \text{constant} \quad (7.6)$$

where the constant may vary from cycle to cycle. The internal stress distribution must balance the pressure force across the wall, that is,

$$\int_{-h/2}^{h/2} \sigma \, dx = \sigma_p h \quad (7.7)$$

Creep Ratchetting

Now if we assume that there is no work hardening upon plastic yielding (perfect plasticity), the yield criterion becomes

$$|\sigma| = \sigma_y \quad \text{in plastic regions}$$

$$|\sigma| < \sigma_y \quad \text{in elastic regions} \tag{7.8}$$

where σ_y is the yield stress. Since the stress cannot exceed yield, plastic strains are induced in the plastic regions to maintain the stress at yield. The uniaxial stress-strain relationship is now written as

$$\varepsilon = \frac{\sigma}{E} + \alpha T_1 + \eta \tag{7.9a}$$

where σ/E is the elastic component, αT_1 is the thermal component, and η is the plastic component. If we substitute T_1, the stress-strain relationship for the first half of each cycle (start-up) is, in view of Eqs. (7.3), (7.4b), and (7.5a),

$$E\varepsilon = \sigma - \frac{2x\sigma_t}{h} + E\eta$$

and for the second half of each cycle (shut-down), when the temperature is removed,

$$E\varepsilon = \sigma + E\eta \tag{7.9c}$$

The problem is to find the stresses and strains with Eqs. (7.9c) or (7.9b) during each cycle subject to the conditions (7.6), (7.7), and (7.8). Certain features of the solution may now be observed. For the first half of each cycle Eq. (7.9b) holds, and since ε is to be independent of x by condition (7.6), we conclude that the combination of $\sigma + E\eta$ must be linear in x with slope $2\sigma_t/h$. Thus in regions where $E\eta$ is uniform σ has a slope $2\sigma_t/h$, and vice versa. For the second half of each cycle Eq. (7.9c) holds and since ε must be uniform, $\sigma + E\eta$ must also be uniform across the wall. Thus for the second half of each cycle σ and $E\eta$ have equal and opposite slopes at each point.

The greatest stress in the tube occurs at $x = h/2$ according to Eq. (7.5). If we limit this to yield, we find that

$$\sigma_p + \sigma_t \leq \sigma_y \tag{7.10}$$

This establishes combinations of elastic behavior; that is, any combination of σ_p and σ_t that is less than σ_y will be elastic, while any combination equal to σ_y will be plastic. Combinations that exceed σ_y will lead to various types of plastic behavior, as outlined in the Introduction. These will now be analyzed following Burgreen's explanation [8] of Bree's original development [5].

Shakedown. Shakedown of the tube can occur in two ways. One way is that the application of the temperature yields only one side of the tube wall, while the second way involves yielding of both sides of the wall. In both cases when the temperature is removed and on all later applications and removals, the behavior will be elastic. We will consider the one sided yielding case in detail.

The stress distribution for yielding on one side on first application of the temperature is given in Fig. 7.1. The yielded zone according to Eq. (7.5) is on the outer part of the wall. The elastic region $-h/2 \leqslant x \leqslant a$, where a is to be determined, has a strain distribution

$$E\varepsilon_1 = \sigma_1 - \sigma_t 2x/h \qquad (7.11a)$$

where the subscript 1 pertains to the first half-cycle. The plastic region $a \leqslant x \leqslant h/2$ has the strain distribution

$$E\varepsilon_1 = \sigma_y - \frac{2x\sigma_t}{h} + E\eta \qquad (7.11b)$$

At $x = a$, continuity requires that $\sigma_1 = \sigma_y$. Thus Eq. (7.11a) gives

$$E\varepsilon_1 = \sigma_y - \frac{\sigma_t 2a}{h} \qquad (7.11c)$$

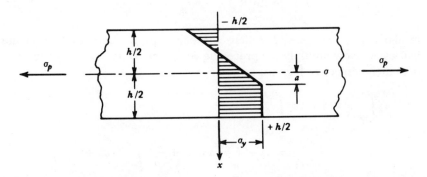

Fig. 7.1 First half cycle stresses with yielding on one side only (S_1). Courtesy of C. P. Press.

and the stress distributions in the two regions are

$$\sigma_1 = \sigma_y \quad a \leq x \leq \frac{h}{2} \quad \text{(plastic)}$$

$$\sigma_1 = \sigma_y + 2\sigma_t \frac{x-a}{h} \quad \frac{-h}{2} \leq x \leq a \quad \text{(elastic)} \tag{7.11d}$$

The plastic strain can be found by substituting $E\varepsilon_1$ from Eq. (7.11c) into Eq. (7.11b), with the result

$$E\eta = 2\sigma_t \frac{x-a}{h} \tag{7.11e}$$

Finally, the elastic-plastic interface location a is found from the equilibrium condition Eq. (7.7) as

$$\frac{a}{h} = \left(\frac{\sigma_y - \sigma_p}{\sigma_t}\right)^{1/2} - \frac{1}{2} \tag{7.11f}$$

With this expression for a the relationships for ε_1, η, and the stress distributions can be obtained by substituting it into Eqs. (7.11c, d, e).

To produce shakedown in these circumstances the stresses obtained by removing the temperature in the second half-cycle must not exceed yield. The changes in the strain after removal of the temperature are

$$E\Delta\varepsilon_2 = \Delta\sigma_2 + 2\sigma_t \frac{x}{h} = \sigma_2 - \sigma_1 + 2\sigma_t \frac{x}{h} \tag{7.12}$$

in the region $-h/2 \leq x \leq a$ and

$$E\Delta\varepsilon_2 = \Delta\sigma_2 + \frac{2\sigma_t x}{h} = \sigma_2 - \sigma_y + \frac{2\sigma_t x}{h} \tag{7.12a}$$

in the region $a \leq x \leq h/2$, where the subscript 2 refers to the second half-cycle. Now multiply the preceding equations by dx and integrate them over their respective regions, as

$$E\Delta\varepsilon_2 \int_{-h/2}^{a} dx + E\Delta\varepsilon_2 \int_{a}^{h/2} dx$$

$$= \int_{-h/2}^{a} \Delta\sigma_2 \, dx + \int_{a}^{h/2} \Delta\sigma_2 \, dx + \frac{2\sigma_t}{h} \int_{-h/2}^{a} dx + \frac{2\sigma_t}{h} \int_{a}^{h/2} dx$$

$$\tag{7.12b}$$

If we examine this expression, we note first of all that the sum of the integrals involving $\Delta\sigma_2$ must be zero since there is no change in the external load during the second half-cycle. Next we observe that the integrals involving σ_t cancel each other. As a result we may conclude that the strain increment $\Delta\varepsilon_2 = 0$. With this result and Eq. (7.11d) for σ_1, Eqs. (7.12) and (7.12a) give

$$\sigma_2 = \sigma_y - \frac{2\sigma_t a}{h} \qquad -h/2 \leqslant x \leqslant a$$

$$\sigma_2 = \sigma_y - \frac{2\sigma_t x}{h} \qquad a \leqslant x \leqslant h/2 \qquad (7.12c)$$

If we re-apply the thermal stress, no further plasticity occurs. We have obtained the limit of elastic behavior, that is, Eq. (7.10), previously. Now let us find the limit of shakedown on one side of the tube wall, as follows.

To avoid yielding on the surface at the end of the second half-cycle we set σ_2 just equal to the yield stress in Eqs. (7.12c). This gives the two conditions

$$\sigma_y - \sigma_t \geqslant -\sigma_y \qquad \text{or} \qquad \sigma_t \leqslant 2\sigma_y$$

$$\sigma_y - \frac{2\sigma_t a}{h} \leqslant \sigma_y \qquad \text{or} \qquad a \geqslant 0 \qquad (7.12d)$$

The second requirement that a be non-negative gives, from Eq. (7.11f),

$$\sigma_p + \frac{\sigma_t}{4} \leqslant \sigma_y \qquad (7.12e)$$

As a last requirement, we must make certain that yielding takes place at the outer surface. Thus a must be less than $h/2$. Again using Eq. (7.11f), this gives the inequality.

$$\sigma_p + \sigma_t \leqslant \sigma_y \qquad (7.12f)$$

This, as could be expected, is the same as Eq. (7.10). Thus the region of shakedown for yielding on one side of the wall only, which is called the S_1 region, is bounded by combinations of stress that satisfy Eqs. (7.12d,e,f). This is depicted in Fig. 7.6.

In a similar fashion the initial application of temperature could yield both regions of the tube, as shown in Fig. 7.2. Now there is a second region of plasticity defined by b in Fig. 7.2 (compare to Fig. 7.1). An analysis

158 Creep Ratchetting

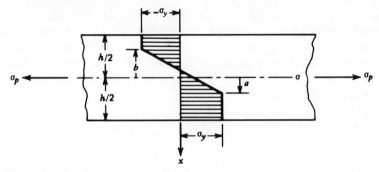

Fig. 7.2 First half cycle stresses with yielding on both sides (S_2). Courtesy of C. P. Press.

similar to that previously given for one-sided yielding now leads to [5], [8]

$$\frac{a}{h} = \frac{1}{2}\left(\frac{\sigma_y}{\sigma_t} - \frac{\sigma_p}{\sigma_y}\right)$$

$$\frac{b}{h} = -\frac{1}{2}\left(\frac{\sigma_y}{\sigma_t} + \frac{\sigma_p}{\sigma_y}\right) \qquad (7.13)$$

The combinations of stress that define shakedown due to two-sided yielding, which is called the S_2 region, are found to be [5], [8]

$$\sigma_t \leq 2\sigma_y$$

$$\sigma_t(\sigma_y - \sigma_p) \geq \sigma_y^2 \qquad (7.13a)$$

These are depicted in Fig. 7.6. Since the first of Eq. (7.13a) is identical to the first of Eq. (7.12d), we note that one-sided yielding is superceded by two-sided yielding in the S_2 region.

Alternating Plasticity. Alternating plasticity (plastic cycling) can occur if the stresses behave as shown in Fig. 7.3. At the end of the first half-cycle the stress distribution is like that of the case of shakedown for two-sided yielding shown in Fig. 7.2. Again a and b are as defined by Eqs. (7.13). However, subsequent behavior is different, as will now be shown. Fig. 7.3b shows the stress change during the second half-cycle, and Fig. 7.3c shows the stress at the end of the second half-cycle. The salient feature of the behavior now is that the central portion of the wall remains elastic. Furthermore, yielding in the outer regions of the tube wall occurs alternately in tension and compression, as shown in Fig. 7.3. Since there is no

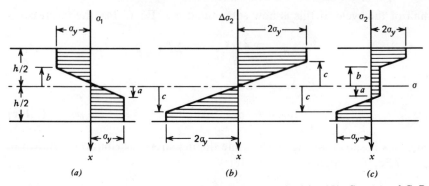

Fig. 7.3 First and second half cycle stress for alternating plasticity (P). Courtesy of C. P. Press.

change in the pressure, the incremental stress $\Delta\sigma_2$ must be self-equilibrating, that is, it should produce no change in the in-plane load. The incremental stress distribution must, therefore, be as shown in Fig. 7.3b, with elastic-plastic boundaries at $x = \pm c$.

In the central portion $-c \leqslant x \leqslant c$ the wall is elastic and the increments of stress and strain are related by

$$E\Delta\varepsilon_2 = \Delta\sigma_2 + \frac{2\sigma_t x}{h} \qquad (7.14)$$

At the elastic-plastic interface $x = -c$, the stress increment is $2\sigma_y$, while at $x = c$ the stress increment is $-2\sigma_y$. If we substitute these into Eq. (7.14), we obtain

$$E\Delta\varepsilon_2 = -2\sigma_y + \frac{2\sigma_t c}{h}$$

$$E\Delta\varepsilon_2 = 2\sigma_y - \frac{2\sigma_t c}{h} \qquad (7.14\text{a})$$

Since the strain change is uniform, subtraction of these equations gives [5], [8]

$$c = \frac{\sigma_y}{\sigma_t} h \qquad (7.14\text{b})$$

In addition, if we now substitute c into either of Eqs. (7.14a), we find that $\Delta\varepsilon_2 = 0$. This means there is no net strain increment during the second half of the cycle. If this is now substituted into Eq. (7.14), the result is

$$\Delta\sigma_2 = -\frac{2\sigma_t x}{h} \quad -c \leqslant x \leqslant c$$

$$\Delta\sigma_2 = -2\sigma_y \quad x \geqslant c$$

$$\Delta\sigma_2 = +2\sigma_y \quad x \leqslant -c \quad (7.14c)$$

and this is shown in Fig. 7.3b, while the net stress distribution is shown in Fig. 7.3c. The latter is given by [8]

$$\sigma_2 = \sigma_y - \frac{2\sigma_t a}{h} \quad -b \leqslant x \leqslant a$$

$$\sigma_2 = \sigma_y - \frac{2\sigma_t x}{h} \quad a \leqslant x \leqslant c$$

$$\sigma_2 = -\sigma_y \quad x \geqslant c$$

$$\sigma_2 = -\sigma_y - \frac{2\sigma_t x}{h} \quad -c \leqslant x \leqslant -b$$

$$\sigma_2 = \sigma_y \quad x \leqslant -c \quad (7.14d)$$

Now the stress strain relationships in the two plastic regions are, with $\Delta\varepsilon_2 = 0$,

$$E\Delta\varepsilon_2 = 0 = -2\sigma_y + \frac{2\sigma_t x}{h} + E\Delta\eta_2 \quad h/2 \geqslant x \geqslant c$$

$$E\Delta\varepsilon_2 = 0 = 2\sigma_y + \frac{2\sigma_t x}{h} + E\Delta\eta_2 \quad -h/2 \leqslant x \leqslant c \quad (7.14e)$$

These can be solved for the plastic strain increments to give [8]

$$E\Delta\eta_2 = 2\sigma_y - \frac{2\sigma_t x}{h} \quad h/2 \geqslant x \geqslant c$$

$$E\Delta\eta_2 = -2\sigma_y - \frac{2\sigma_t x}{h} \quad -h/2 \leqslant x \leqslant c \quad (7.14f)$$

When the temperature is re-applied in the third half-cycle, the incremental stress distribution is the reverse of that shown in Fig. 7.3b. If this is added to the stress distribution of Fig. 7.3c, we get the distribution of Fig. 7.3a back. Subsequent additions and removals of temperature cause the stress distribution to fluctuate between Fig. 7.3a and Fig. 7.3c.

A Thin Tube Under Internal Pressure and Cyclic Thermal Stresses

Combinations of stresses that produce alternating plasticity as just described and that define the so-called P region are now bounded as follows. We first require that $|c| < h/2$, otherwise, the increments would be elastic and there would be no alternating plasticity. From Eq. (7.14b) this gives

$$\sigma_t \geqslant 2\sigma_y \qquad (7.14\text{g})$$

The second condition is obtained by noting that the stress is the central core of the wall as shown in Fig. 7.3c must always be no greater than σ_y. The first of Eqs. (7.14d) thus gives

$$a \geqslant 0 \qquad (7.14\text{h})$$

and the first of Eqs. (7.13) yields the result [5], [8]

$$\sigma_p \sigma_t \leqslant \sigma_y^2 \qquad (7.14\text{i})$$

The boundaries Eqs. (7.14g, i) are shown in Fig. 7.6, and define the P region of alternating plasticity.

Ratchetting. The final case to be considered is ratchetting. As in shakedown, there are two possibilities: ratchetting due to yielding on one side of the tube wall, called R_1, or ratchetting due to yielding on both sides of the tube wall, called R_2. We will derive conditions for the one-sided case, referring to shakedown for one-sided yielding as previously derived here. We may recall in that analysis we found by Eq. (7.12d) that a, the boundary between elastic and plastic behavior had to be positive. Now we relax this requirement and permit the boundary to extend past the midwall of the tube, as shown in Fig. 7.4a. The first half-cycle analysis is

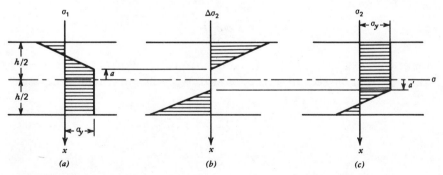

Fig. 7.4 First and second half cycle stresses for ratchetting with yielding on one side only (R_1). Courtesy of C. P. Press.

the same as for the S_1 shakedown analysis, so the expressions for the stresses, strains, and a remain the same.

For the second half-cycle the stress-strain relations take the form

$$E\Delta\varepsilon_2 = \Delta\sigma_2 + \frac{2\sigma_t x}{h} + E\Delta\eta_2$$

$$= \sigma_y - \sigma_1 + \frac{2\sigma_t x}{h} + E\Delta\eta_2 \qquad -\frac{h}{2} \leq x \leq a$$

$$E\Delta\varepsilon_2 = \frac{2\sigma_t x}{h} + E\Delta\eta_2 \qquad a \leq x \leq a'$$

$$E\Delta\varepsilon_2 = \Delta\sigma_2 + \frac{2\sigma_t x}{h}$$

$$= \sigma_2 - \sigma_y + \frac{2\sigma_t x}{h} \qquad a' \leq x \leq \frac{h}{2} \qquad (7.15)$$

The stress distributions are shown in Fig. 7.4. We assume that on removal of the temperature the distribution of Fig. 7.4a becomes that of Fig. 7.4c. The loading is beyond the shakedown range since a extends across the midwall. As stipulated at the outset, the strain increment $\Delta\varepsilon_2$ has to be uniform across the wall. If we apply the third of Eqs. (7.15) at $x = a'$, where $\sigma_2 = \sigma_y$, we obtain

$$E\Delta\varepsilon_2 = \frac{2\sigma_t a'}{h} \qquad (7.15a)$$

With this, the stress increment in the third of Eqs. (7.15) becomes

$$\Delta\sigma_2 = \frac{2\sigma_t(a'-x)}{h}$$

$$\sigma_2 = \sigma_y + \Delta\sigma_2 = \sigma_y + \frac{2\sigma_t(a'-x)}{h} \qquad (7.15b)$$

for $a' \leq x \leq h/2$. We can also find $\Delta\sigma_2$ in the first of Eqs. (7.15) because we know σ_1 from the second of Eqs. (7.11d). With this we obtain

$$\Delta\sigma_2 = \sigma_y - \sigma_1 = \frac{2\sigma_t(a-x)}{h}$$

$$\sigma_2 = \sigma_y \qquad (7.15c)$$

for $-h/2 \leq x \leq a$. We may now find a and a' by imposing the equilibrium

A Thin Tube Under Internal Pressure and Cyclic Thermal Stresses

condition of Eq. (7.7) to the increments, noting that these take place without any change in the external load. Hence

$$\int_{-h/2}^{a} \Delta\sigma_2 \, dx + \int_{a'}^{h/2} \Delta\sigma_2 \, dx = 0 \qquad (7.15d)$$

because $\Delta\sigma_2$ is zero for $a \leq x \leq a'$. If we substitute the required information from Eqs. (7.15b,c) into the foregoing, we find

$$\frac{a'}{h} = -\frac{a}{h} = \frac{1}{2} - \left(\frac{\sigma_y - \sigma_p}{\sigma_t}\right)^{1/2} \qquad (7.16d)$$

Note that a is identical to the value we obtained previously for the shakedown calculation, as in Eq. (7.11f).

Now let us calculate the plastic strain increments in the second half-cycle. From Eqs. (7.15) these are

$$E\Delta\eta_2 = \frac{4\sigma_t a'}{h} \qquad \frac{-h}{2} \leq x \leq a$$

$$E\Delta\eta_2 = \frac{2\sigma_t(a'-x)}{h} \qquad a \leq x \leq a' \qquad (7.16e)$$

In the third half-cycle the temperature is re-applied. Starting with the stress distribution in Fig. 7.5a (which is identical to that of Fig. 7.4c) the addition of the temperature gradient produces the stress changes $\Delta\sigma_3$ shown in Fig. 7.5b and the final distribution of σ_3 shown in Fig. 7.5c (which is identical to σ_1 in Fig. 7.4, as will be shown).

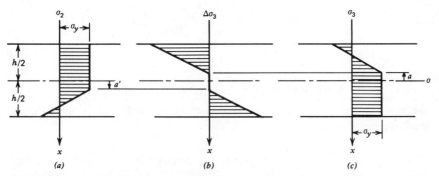

Fig. 7.5 Second and third half cycle stresses for ratchetting with initial yielding on one side only (R_1). Courtesy of C. P. Press.

The stress-strain relationships for the three regions are now

$$E\Delta\varepsilon_3 = \Delta\sigma_3 - \frac{2\sigma_t x}{h}$$

$$= \sigma_3 - \sigma_y - \frac{2\sigma_t x}{h} \qquad \frac{-h}{2} \leqslant x \leqslant a$$

$$E\Delta\varepsilon_3 = -\frac{2\sigma_t x}{h} + E\Delta\eta_3 \qquad a \leqslant x \leqslant a'$$

$$E\Delta\varepsilon_3 = \Delta\sigma_3 - \frac{2\sigma_t x}{h} + E\Delta\eta_3$$

$$= \sigma_y - \sigma_2 - \frac{2\sigma_t x}{h} + E\Delta\eta_3 \qquad a' \leqslant x \leqslant \frac{h}{2} \qquad (7.16f)$$

We can find the uniform strain from the first of these by noting that $\sigma_3 = \sigma_y$ at $x = a$. This gives

$$E\Delta\varepsilon_3 = \frac{-2\sigma_t a}{h} \qquad (7.16g)$$

and the stresses become

$$\Delta\sigma_3 = \frac{2\sigma_t(x-a)}{h}$$

$$\sigma_3 = \sigma_y + \Delta\sigma_3 = \sigma_y + 2\sigma_t(x-a) \qquad (7.16h)$$

for $-\frac{h}{2} \leqslant x \leqslant a$. Similarly, Eq. (7.12c) can be used to find

$$\Delta\sigma_3 = \sigma_y - \sigma_2 = \frac{-2\sigma_t(a'-x)}{h}$$

$$\sigma_3 = \sigma_y \qquad (7.16i)$$

for $a' \leqslant x \leqslant \frac{h}{2}$. If we now enforce the equilibrium condition of Eq. (7.7) on the strain increment $\Delta\sigma_3$, and note that there is no change in the external load, we again obtain

$$\int_{-h/2}^{a} \Delta\sigma_3 \, dx + \int_{a'}^{h/2} \Delta\sigma_3 \, dx = 0 \qquad (7.16j)$$

which yields the same result as before, namely,

$$\frac{a'}{h} = -\frac{a}{h} = \frac{1}{2} - \left(\frac{\sigma_y - \sigma_p}{\sigma_t}\right)^{1/2} \qquad (7.16\text{k})$$

This verifies that we get the original stress pattern after the third half-cycle. Finally, the plastic strain increments can be found from Eq. (7.16g, f) as

$$E\Delta\eta_3 = \frac{2\sigma_t(x-a)}{h} \qquad a \leqslant x \leqslant a'$$

$$E\Delta\eta_3 = \frac{2\sigma_t(a'-a)}{h} \qquad a' \leqslant x \leqslant \frac{h}{2} \qquad (7.16\text{l})$$

The net plastic strain developed in a full cycle, or the cyclic ratchetting growth, is obtained by adding the plastic strain increments in the second and third half-cycles. Thus we add the plastic strain increments for the region $a \leqslant x \leqslant a'$ from Eqs. (7.16e,l) to give the growth as

$$E\varepsilon_g = \frac{4\sigma_t a'}{h} \qquad (7.16\text{m})$$

and on substitution of Eq. (7.16b) this becomes

$$E\varepsilon_g = 4\sigma_t \left[\frac{1}{2} - \left(\frac{\sigma_y - \sigma_p}{\sigma_t}\right)^{1/2}\right] \qquad (7.16\text{n})$$

Finally, we must determine the bounds for the combinations of stress that give this type of ratchetting. The first requirements is that $\varepsilon_g \geqslant 0$. Eq. (7.16n) therefore gives

$$\sigma_p + \frac{\sigma_t}{4} \geqslant \sigma_y \qquad (7.16\text{o})$$

The second requirement is that there be plastic yielding only on one side of the tube wall. Therefore, σ_2 or σ_3 must each be greater than $-\sigma_y$ at $x = h/2$ and $x = -h/2$, respectively. This leads to

$$\sigma_t(\sigma_y - \sigma_p) \leqslant \sigma_y^2 \qquad (7.16\text{p})$$

The final requirement is that there be no plastic collapse due to σ_p alone, that is,

$$\sigma_p \leqslant \sigma_y \qquad (7.16\text{q})$$

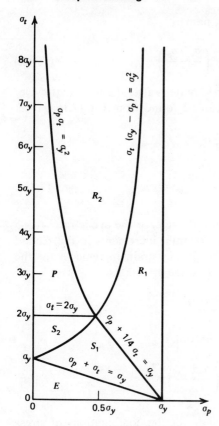

Fig. 7.6 Interaction diagram. Region definition: E = elastic cycling, P = plastic cycling, S_1 and S_2 = shakedown after first half cycle, R_1 and R_2 = ratchetting [5].

The three boundaries Eqs. (7.16o,p,q) define what is known as the R_1 region and are shown in Fig. 7.6.

In a similar fashion the initial application of temperature can cause a second zone of yielding on the inner surface, as shown in Fig. 7.2. In this case a and b, as defined by Fig. 7.2, are again shown [5], [8] to be given by Eq. (7.13). The growth in this case is shown to be

$$E\varepsilon_g = 2\sigma_t\left(\frac{\sigma_p}{\sigma_y} - \frac{\sigma_y}{\sigma_t}\right) \qquad (7.17)$$

and the boundaries for this type of ratchetting are obtained as

$$\sigma_p\sigma_t > \sigma_y^2$$

$$\sigma_t(\sigma_y - \sigma_p) > \sigma_y^2 \qquad (7.17a)$$

A Thin Tube Under Internal Pressure and Cyclic Thermal Stresses 167

The boundaries Eqs. (7.17a) define what is called the R_2 region, as shown in Fig. 7.6.

All of the behavior that has just been described is brought together in Fig. 7.6, the so-called Bree diagram [5]. To determine which type of behavior occurs in a given situation, one need only calculate σ_p and σ_t according to Eqs. (7.2) and (7.4b), and plot the point in Fig. 7.6. In studying that figure we may note that while it is bounded on the right by the collapse load $\sigma_p = \sigma_y$, there is no such limit on the thermal stress. This is because thermal stress alone cannot cause collapse. Note also that if there is no mechanical load, ratchetting will not occur. Similarly, if there is no thermal load, no plastic behavior of any type can occur except collapse. Finally, we should emphasize that this is a one-dimensional approximation that was conceived primarily to permit closed form solutions that bring out the essential features of the behavior. Although we have developed the solution for the case of perfect plasticity, the behavior is not subject to this restriction. Miller [1] and Bree [5] showed that it can also occur when there is work hardening.

7.2.2 Effect of Stress Relaxation due to Creep

The foregoing results are valid if the mean temperature of the tube is low enough for creep to be ignored. However, it is reasonable to assume that during the part of the cycle when the temperature acts, the temperature level should be sufficiently high for stress relaxation due to creep to occur. The question of how much relaxation of stress occurs during the hot part of the cycle now arises. In one analysis by Bree [5] total relaxation of the thermal stresses was assumed and the resulting behaviors in the various regimes of Fig. 7.6 were re-examined. Later, Bree [7] assumed only partial relaxation of stress and repeated the analysis. For simplicity we present the complete relaxation analysis as it was derived by Bree [5] and later by Burgreen [8]. For brevity we examine only the S_1 region and then give the results for the other regions of Fig. 7.6 without derivation.

As before, the first application of temperature for the S_1 region appears as Fig. 7.1, where the yield stress is now taken as that at the upper temperature. The stresses, total strains, and plastic strains in the various regions are again as given by Eqs. (7.11a, b, c, d, e) and the elastic-plastic interface a is given by Eq. (7.11f).

At this point the stresses in the tube begin to relax. It is assumed that the thermal stresses relax completely, and that the stress becomes uniform and equal to σ_p across the wall. The tube wall also undergoes creep deformation due to σ_p, but we are not dealing with this at this time. The cyclic growth is what we are interested in and this is computed as follows:

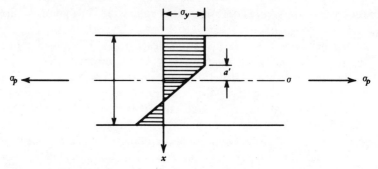

Fig. 7.7 Stress distribution for second half cycle after relaxation and removal of temperature. Courtesy of C. P. Press.

When the temperature is removed after relaxation takes place, the net stress distribution is as shown in Fig. 7.7. The stress-strain relation is

$$E\Delta\varepsilon_2 = \sigma_2 - \sigma_p + \frac{2\sigma_t x}{h} \tag{7.18}$$

Since the stress at $x = a'$ is σ_y, the foregoing equation leads to

$$E\Delta\varepsilon_2 = \sigma_y - \sigma_p + \frac{2\sigma_t a'}{h}$$

$$\sigma_2 = \sigma_y + \frac{2\sigma_t(a'-x)}{h} \qquad a' \leqslant x \leqslant \frac{h}{2}$$

$$\sigma_2 = \sigma_y \qquad \frac{-h}{2} \leqslant x \leqslant a' \tag{7.18a}$$

As before, the location of a' is obtained from the equilibrium Eq. (7.7) to be

$$\frac{a'}{h} = \frac{1}{2} - \left(\frac{\sigma_y - \sigma_p}{\sigma_t}\right)^{1/2} = -\frac{a}{h} \tag{7.18b}$$

When this is substituted into the first of Eqs. (7.18a) the strain increment becomes

$$E\Delta\varepsilon_2 = \sigma_y - \sigma_p + \sigma_t\left[1 - 2\left(\frac{\sigma_y - \sigma_p}{\sigma_t}\right)^{1/2}\right] \tag{7.18c}$$

On re-application of the temperature it is assumed that a completely elastic stress distribution occurs, because we wish to remain in the S_1 region. The elastic stress increment is then

$$\Delta\sigma_3 = \frac{2\sigma_t x}{h} \qquad (7.18d)$$

and on addition of this to σ_2 in Eqs. (7.18a) we obtain

$$\sigma_3 = \sigma_y + \frac{2\sigma_t a'}{h} \qquad a' \leqslant x \leqslant \frac{h}{2}$$

$$\sigma_3 = \sigma_y + \frac{2\sigma_t x}{h} \qquad \frac{-h}{2} \leqslant x \leqslant a' \qquad (7.18e)$$

There is no strain increment, that is, $\Delta\varepsilon_3 = 0$ because the elastic stress increment $\Delta\sigma_3$ is antisymmetric about the midwall of the tube. If the cycling now continues, the stresses and strains just calculated for the second and third half-cycles repeat themselves. The net cyclic growth is due solely to $\Delta\varepsilon_2$ and thus [5], [8]

$$E\varepsilon_g = \sigma_y - \sigma_p + \sigma_t \left\{ 1 - 2\left(\frac{\sigma_y - \sigma_p}{\sigma_t}\right)^{1/2} \right\} \qquad (7.18f)$$

for the S_1 region. It has been further shown by similar analyses that, with complete relaxation of the thermal stress during the hot part of the cycle, the growth is [5]

$$E\varepsilon_g = \sigma_y - \sigma_p + 2\sigma_t \left\{ 1 - 2\left(\frac{\sigma_y - \sigma_p}{\sigma_t}\right)^{1/2} \right\} \qquad (7.18g)$$

in the R_1 region,

$$E\varepsilon_g = \sigma_y - \sigma_p + \sigma_t \left(\frac{\sigma_p}{\sigma_y} - \frac{\sigma_y}{\sigma_t}\right) \qquad (7.18h)$$

in the S_2 and P regions, and

$$E\varepsilon_g = \sigma_y - \sigma_p + 2\sigma_t \left(\frac{\sigma_p}{\sigma_y} - \frac{\sigma_y}{\sigma_t}\right) \qquad (7.18i)$$

in the R_2 region. Thus cyclic growth (ratchetting) now occurs in every region of Fig. 7.6 except the E region. Moreover, where ratchetting previously occurred in the R_1 and R_2 regions the cyclic growth with stress relaxation in those regions is increased by $(\sigma_y - \sigma_p)/E$ per cycle. *Note that this is in addition to the creep strain developed by the pressure stresses when the temperature is on.* We have thus discovered a rather severe result; namely, that if there is relaxation due to creep, there will be ratchetting, now called creep ratchetting, whenever

$$\sigma_t + \sigma_p > \sigma_y \quad (7.17j)$$

This criterion is valid even if the relaxation is partial [7]. It states that only elastic combinations of the thermal and pressure stresses will experience no creep ratchetting.

7.2.3 A Bounding Technique for Strains due to Creep Ratchetting

If we examine Figs. 7.1, 7.2, and 7.3 we will observe that in each of these, which describe the behavior in the S_1, S_2, and P regions, there is always an elastic core in the wall, with elastic-plastic interfaces a, b, and c as given by Eqs. (7.11f), (7.13), and (7.14b). O'Donnell and Porowski [9] found from these diagrams that the elastic core stresses could be expressed as

$$\sigma_c = \sigma_t + \sigma_y - 2\left[\sigma_t(\sigma_y - \sigma_p)\right]^{1/2} \quad (7.19)$$

for region S_1, and

$$\sigma_c = \frac{\sigma_p \sigma_t}{\sigma_y} \quad (7.19a)$$

for regions S_2 and P.

Using these equations O'Donnell and Porowski [9] constructed lines of constant σ_c for the S_1, S_2, and P regions of Fig. 7.6. These are shown in Fig. 7.8. Each line represents combinations of σ_p and σ_t that give constant values of σ_c. This constant elastic core stress represents a way to obtain an upper bound on the creep strains accumulated over the life of the tube. From the elastic core stress and its duration of application, one can obtain the resulting strain from an isochronous stress-strain curve for the material and temperature that are involved.

The maximum creep strains that occur during any loading cycle can be calculated by using the previous technique since it includes the increase in stress in the elastic core that is caused by the alternating stresses. Hence, as stated, the time during which the stress and the temperature are in the creep range during the cycle can be used to bound the creep strain $\Delta\varepsilon_{CR}$ as described, using the core stress from Fig. 7.8. There is also a relaxation strain increment $\Delta\varepsilon_R$ that can be found by considering the elastically calculated peak stress that occurs during the time t_c for which the peak

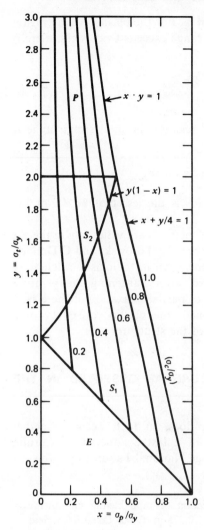

Fig. 7.8 Interaction diagram with lines of constant core stress [9].

stress and temperature are in the creep range. If we use a power law for steady creep, then according to the methods of Chapter 3 the stress after relaxation during time t_c is (compare with Eq. (3.29):

$$|\sigma_r| = \left((|\sigma_{\text{peak}}|)^{1-m} + Ea(m-1)t_c\right)^{1/(1-m)} \quad (7.20)$$

The strain range due to the relaxation of the stress is then

$$\Delta\varepsilon_R = \frac{\sigma_{\text{peak}} - \sigma_r}{E} \quad (7.20a)$$

Note that σ_{peak} is a combination of σ_p and σ_t. For the plastic cycling region P, where the elastically calculated stress range exceeds twice yield, there is also plastic cycling $\Delta\varepsilon_p$. This is equal to

$$\Delta\varepsilon_P = \frac{|\sigma_t| - 2\sigma_y}{E} \tag{7.20b}$$

because there is no ratchetting in the P region. The total strain range is then

$$\Delta\varepsilon = \Delta\varepsilon_{CR} + \Delta\varepsilon_R + \Delta\varepsilon_P + \Delta\varepsilon_E$$

where $\Delta\varepsilon_E$ is the elastic strain range. This is the lesser of $2\sigma_y/E$ or the elastically calculated local strain range.

It should be emphasized that the above methods are valid only in the S_1, S_2, and P regions of Figs. 7.6 and 7.8. Recently, Porowski and O'Donnell [10] extended their approach to permit operation in the plastic ratchetting R_1 and R_2 regions. They also included work hardening, temperature dependent yield properties, and the idea of partial relaxation of the elastic core stress. In the method described here the elastic core stress was maintained at its full value over the life of the structure.

7.3 CREEP RATCHETTING OF A TUBE AS OBSERVED IN THE LABORATORY

To demonstrate that creep ratchetting does, in fact, occur in actual equipment we now present the results of a test conducted by Corum, Young, and Grindell [11] at the Oak Ridge National Laboratory. We start with a brief description of the test:

> The test specimen was a 30-in.-long section of pipe having an outside diameter of 8.44 in. and a wall thickness of 0.375 in. It was taken from Heat 9T2796 of type 304 stainless steel. The specimen was welded, after final annealing, to adjacent extensions of 8-in. stainless steel pipe of the same dimensions. The annealing heat treatment consisted of heating the specimens to 2000°F for 1/2 hour and then forced-air cooling rapidly to room temperature.
>
> The *nominal* temperature and pressure histories of the liquid sodium coolant for the test are depicted in Fig. 7.9. The nominal sodium temperature and pressure were 1100°F and 700 psi respectively during the long-term hold periods. The thermal down shock was from 1100° to 800°F at a *nominal* rate of 30°F/sec. Near adiabatic conditions were maintained on the outer surface of the specimen during each thermal transient. At 800°F, the internal

pressure was removed, reapplied, and the temperature was slowly returned to 1100°F at a rate of 50°F/hr. The specimen was then held at 1100°F and subjected to the internal pressure of 700 psi for a period of 160 hrs. before the next transient was initiated. The total cycle time was 168 hrs. (1 week), with approximately 2 hrs. spent at 800°F and 6 hrs. spent in heating from 800° to 1100°F. The specimen was subjected to a total of 13 cycles.

The idealized ramp transient shown in [Fig. 7.9] was not actually obtained in the tests. The measured sodium thermal transients imposed on the pipe specimen were more realistic and are shown in [Fig. 7.10]. These transients were measured at the center of the test sections. They were reproducible from cycle to cycle and test to test. There was a slight temperature gradient along the specimen axis during the actual transients (about 1.3°F/in. in the most severe transient). It is believed that this gradient, which was approximately linear, can be neglected in any analyses of the problem.

The test facility in which the specimens were tested was designed to virtually eliminate piping reactions in the pipe ratchetting test assemblies, and measurements of strains during heatup of the system indicated that if any net reactions did exist, they were very small. Thus, each test specimen can be analyzed as an infinitely long straight pipe with only the internal pressure loading (due to the pressure acting of the inside area of the specimen) in the axial direction.

The measured ratchetting behavior for the first pipe specimen is shown in [Fig. 7.11], where the circumferential strain on the outer surface is plotted as a function of the accumulated hold time at 1100°F. Data points from two capacitive gages located 90 deg apart near the midlength of the 30-in. test section are shown. Initial pressurization data and data corresponding to the minimum strain during each transient are included.

The response of the specimen indicates substantial ratchetting in the early cycles, but the incremental growth per cycle decreased continually with an increasing number of cycles. The net strains from the two circumferential capacitive gages on the specimen agree reasonably well with each other although there does seem to be a consistent difference [gage 513 generally indicates more plastic (time-independent) strain and less creep (time-dependent) strain than does gage 514]. These differences are thought to be real. A possible explanation is the significant variation in properties that has been found around the circumference of the 8-in. pipe product from which ratchetting specimens were obtained. Also, although extreme care was taken to support the specimens so that negligible end reactions existed, it is possible that very small reactions did reach the specimens. Although small, these could have a significant influence, particularly during creep periods.

The solid lines in Fig. 7.11 are results that were computed by the investigators. These are included to show the accuracy that can be obtained with digital computer solutions. We shall return to these results in Chapter 8, where we take up the numerical analysis of creep.

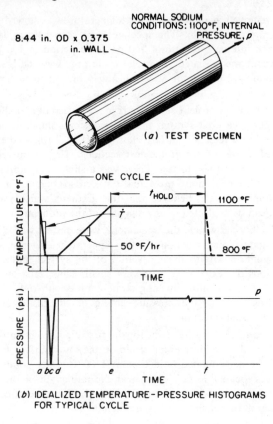

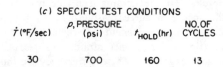

Fig. 7.9 Ratchetting test conditions [11]. Courtesy of the Oak Ridge National Laboratory.

7.4 CLOSURE

In this chapter we have introduced the idea of creep ratchetting by presenting analytical results for a simplified tube problem and experimental results for an actual tube undergoing a typical history of pressure and thermal loading. Generally, the analysis of creep ratchetting of a component is an extremely complex problem that is most effectively handled with a digital computer solution. This is taken up in Chapter 8. Nevertheless, the simplified approach that has been presented here is interesting for two

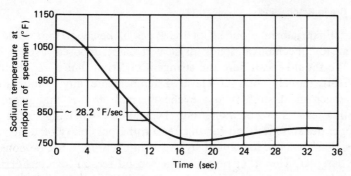

Fig. 7.10 Actual coolant temperature [11]. Courtesy of the Oak Ridge National Laboratory.

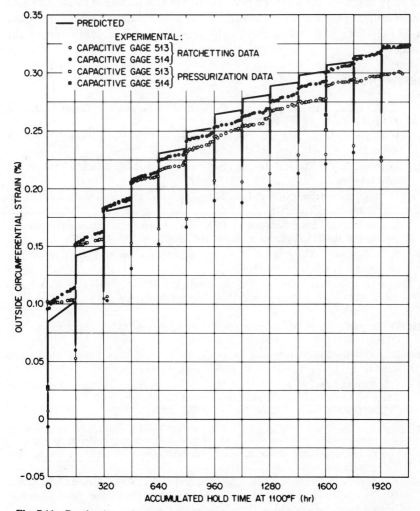

Fig. 7.11 Ratchetting test results [11]. Courtesy of the Oak Ridge National Laboratory.

reasons. First, it brings out all of the types of behavior that could occur when steady mechanical loads exist with superimposed cyclic thermal loads. We should recall here the stringent criterion that is represented by Eq. (7.18j), namely, that at elevated temperature any combination of steady mechanical stresses and cyclic thermal stresses that exceeds the yield stress causes creep ratchetting. Typical elevated temperature equipment operates this way and for this reason the possibility of unacceptable dimensional changes, as caused by creep ratchetting, must be considered by the designer. The latter provides the second reason for considering the simplified approach here. The method of O'Donnell and Porowski [9], which is based on the work of Bree [5], has found its way into the elevated temperature criteria of the A.S.M.E. Boiler and Pressure Vessel Code Case N-47 [12]. This is a recognition of the conservatism and usefulness of the method.

REFERENCES

[1] D. R. Miller, Thermal Stress Ratchet Mechanism in Pressure Vessels, *A.S.M.E. J. Basic Eng.*, vol. 81, pp. 190–196 (1959).

[2] H. G. Edmunds, and F. J. Beer, Notes on Incremental Collapse in Pressure Vessels, *J. Mech. Eng. Sci.*, vol. 5, pp. 187–199 (1961).

[3] D. Burgreen, The Thermal Ratchet Mechanism, *A.S.M.E. J. Basic Eng.*, vol. 90, pp. 319–324 (1968).

[4] D. Burgreen, Structural Growth Induced by Thermal Cycling, *A.S.M.E. J. Basic Eng.*, vol. 90, pp. 469–475 (1968).

[5] J. Bree, Elastic Plastic Behavior of Thin Tubes Subjected to Internal Pressure and Intermittent High Heat Fluxes, *J. Strain Anal.*, vol. 2, pp. 226–238 (1967).

[6] B. E. Gatewood, The Problem of Strain Accumulation Under Thermal Cycling, *J. Aerospace Sci.*, vol. 27, p. 461 (1960).

[7] J. Bree, Incremental Growth Due to Creep and Plastic Yielding of Thin Tubes Subjected to Internal Pressure and Cyclic Thermal Stresses, *J. Strain Anal.*, vol. 3, pp. 122–127 (1968).

[8] D. Burgreen, *Design Methods for Power Plant Structures*, C.P. Press, Jamaica, N.Y. (1975).

[9] W. J. O'Donnell and J. Porowski, Upper Bounds for Accumulated Strains Due to Creep Ratchetting, *A.S.M.E. J. Press. Vess. Tech.*, vol. 96, p. 150 (1974).

[10] J. S. Porowski and W. J. O'Donnell, More Efficient Creep Ratchetting Bounds, Oak Ridge National Laboratory Report on Contract No. W-7405-Eng-26, Subcontract No. 7322, O'Donnell and Associates, Inc., Pittsburgh (1979).

[11] J. M. Corum, H. C. Young, and A. G. Grindell, Thermal Ratchetting in Pipes Subjected to Intermittent Thermal Downshocks at Elevated Temperatures, in *Pressure Vessels and Piping: Verification and Qualification of Inelastic Computer Programs*, A.S.M.E., New York, pp. 79–98 (1975).

[12] American Society of Mechanical Engineers, Boiler and Pressure Vessel Code, Section III, Division 1, Rules for Construction of Nuclear Power Plant Components, Code Case Interpretation N-47, A.S.M.E., New York (1980).

CHAPTER 8

NUMERICAL ANALYSIS OF CREEP

8.1 INTRODUCTION

Throughout this book the solutions that we presented have been subjected to one approximation or another. In most cases it was assumed that the stresses were stationary. This sufficed to bring out the essentials of creep behavior for several important classes of problems, namely, deformation, relaxation, rupture, buckling, and ratchetting. We found it impossible to analyze a problem using any transient creep law except for the time hardening law. The more accurate strain hardening model did not permit closed form solutions to be carried out. In all of the earlier chapters we noted, and deferred to the present chapter, the idea that a numerical solution of the creep problem was the only feasible way to treat the general situation without appealing to an assumption regarding the behavior. Thus we now take up the numerical solution of creep problems, as carried out today with the aid of the digital computer.

We do this by first giving a brief summary of the finite element method, and the manner in which it is modified to permit carrying out creep analyses. This is followed by a review of general purpose computer programs that are currently being used in industry for creep analysis. We end the discussion by presenting some typical solutions that have been made possible by the numerical approach with both general purpose and special purpose programs.

8.2 SUMMARY OF FINITE ELEMENT METHOD

Concerning creep analysis, the predominant numerical approach that has been used is the finite element method. This is probably because, by the time analysts became interested in the creep problem in the early 1970s, the finite element method had emerged as a potent and well developed tool for stress analysis. Therefore, creep analysts turned to it immediately and did not consider some of the earlier techniques for solving differential equations, such as finite differences and direct (stepwise) integration, that have been used in such other areas as shell theory, elasticity, and plasticity.

It is beyond the scope of this book to present a thorough derivation of the finite element method. This is better left to some of the excellent treatments of it that are available today, such as the texts of Zienkiewicz [1], Gallagher [2], Ural [3], and others. We give rather, a brief summary of the method so that we can focus our attention on the manner in which it is adapted to creep analysis.

In the finite element method the structure or continuum under consideration is separated into a collection of a number of so-called finite elements. These elements are connected at nodal points whose displacements become the basic unknown quantities of the problem. This is the so-called displacement formulation. The state of displacement within each element as a function of the nodal displacements is now chosen. These functions also define the state of strain in the elements as a function of the nodal displacements. The strains in turn lead to the stresses within the element through a constitutive law. Finally, a system of forces expressed in terms of the nodal displacements is found. These forces must be in equilibrium with the boundary loads and any distributed load. An outline of the derivation of the method as presented by Zienkiewicz [1] and Ural [3] proceeds as follows: The development is illustrated for plane stress using a triangle as the basic element. In Fig. 8.1 we show one element of the continuum with nodal points i, j, and m. At each node of the triangle there are two displacement components u, and v. Therefore, we have an element displacement vector

$$\{\delta\}^e = \begin{Bmatrix} u_i \\ v_i \\ u_j \\ v_j \\ u_m \\ v_m \end{Bmatrix} \tag{8.1}$$

Summary of Finite Element Method 179

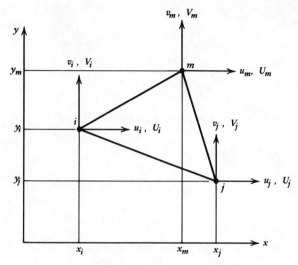

Fig. 8.1 The plane stress triangle element.

Now we assume the displacement distribution in the element as

$$u = \alpha_1 + \alpha_2 x + \alpha_3 y$$
$$v = \alpha_4 + \alpha_5 x + \alpha_6 y \tag{8.2}$$

Through the strain displacement equations this form ensures that the strain is constant over the element. The α_i in the above equations are found by applying them to the nodes. This gives, for example,

$$u_i = \alpha_1 + \alpha_2 x_i + \alpha_3 y_i$$
$$u_j = \alpha_1 + \alpha_2 x_j + \alpha_3 y_j \tag{8.2a}$$
$$u_m = \alpha_1 + \alpha_2 x_m + \alpha_3 y_m$$

and similarly for α_4, α_5, and α_6 from the expression for v. We can solve Eqs. (8.2a) for α_1, α_2, and α_3, and similarly for α_4, α_5, and α_6, by using Cramer's Rule. Then, when we substitute the results back into Eq. (8.2) we obtain

$$u = \frac{1}{2\Delta} \{ (a_i + b_i x + c_i y) u_i + (a_j + b_j x + c_j y) u_j + (a_m + b_m x + c_m y) u_m \} \tag{8.3a}$$

$$v = \frac{1}{2\Delta} \{ (a_i + b_i x + c_i y) v_i + (a_j + b_j x + c_j y) v_j + (a_m + b_m x + c_m y) v_m \} \tag{8.3b}$$

where

$$a_i = x_j y_m - x_m y_j$$
$$b_i = y_j - y_m$$
$$c_i = x_m - x_j \tag{8.3c}$$

and the rest of the coefficients are obtained from Eq. (8.3c) by permutation of subscripts.

In addition, we also have

$$2\Delta = \begin{vmatrix} 1 & x_i & y_i \\ 1 & x_j & y_j \\ 1 & x_m & y_m \end{vmatrix} \tag{8.3d}$$

which can be shown to be twice the area of $\Delta(i, j, m)$. Now let us define

$$N_i = \frac{a_i + b_i x + c_i y}{2\Delta} \tag{8.4a}$$

and so on for N_j and N_m. With these definitions we can write

$$\{\phi\} = \begin{pmatrix} u(x,y) \\ v(x,y) \end{pmatrix} = \begin{Bmatrix} N_i & 0 & N_j & 0 & N_m & 0 \\ 0 & N_i & 0 & N_j & 0 & N_m \end{Bmatrix} \begin{Bmatrix} u_i \\ v_i \\ u_j \\ v_j \\ u_m \\ v_m \end{Bmatrix} \tag{8.4b}$$

or in matrix terms

$$\{\phi\} = [N]\{\delta\}^e \tag{8.4c}$$

The matrix $[N]$ can be viewed as a position matrix. Having thus set up the displacement matrix, we can now discuss the strains. For plane stress we have three nonvanishing strain components

$$\{\varepsilon\} = \begin{Bmatrix} \varepsilon_x \\ \varepsilon_y \\ \gamma_{xy} \end{Bmatrix} = \begin{Bmatrix} \dfrac{\partial u}{\partial x} \\ \dfrac{\partial v}{\partial y} \\ \dfrac{\partial u}{\partial y} + \dfrac{\partial v}{\partial x} \end{Bmatrix} \tag{8.5a}$$

Note that we are using engineering strains in this discussion. We can express the displacement derivatives in terms of Eq. (8.3a,b). Then, when we substitute the result into Eq. (8.5a) we find that

$$\{\varepsilon\} = \frac{1}{2\Delta} \begin{bmatrix} b_i & 0 & b_j & 0 & b_m & 0 \\ 0 & c_i & 0 & c_j & 0 & c_m \\ c_i & b_i & c_j & b_j & c_m & b_m \end{bmatrix} \begin{Bmatrix} u_i \\ v_i \\ u_j \\ v_j \\ u_m \\ v_m \end{Bmatrix} \quad (8.5b)$$

or in the matrix form

$$\{\varepsilon\} = [B]\{\delta\}^e \quad (8.5c)$$

Here the $[B]$ matrix is independent of position because we have defined a constant strain triangle.

With the strains we can now find the stresses. For plane stress there are three nonvanishing stress components that can be written

$$\{\sigma\} = \begin{Bmatrix} \sigma_x \\ \sigma_y \\ \tau_{xy} \end{Bmatrix} \quad (8.6a)$$

To obtain the stresses in terms of the strains we assume for the moment that the continuum is elastic. Thus we use Hooke's law and obtain

$$\{\sigma\} = [D]\{\varepsilon\} \quad (8.6b)$$

where $[D]$ is called the elasticity matrix. For plane stress we can show that it takes the form

$$[D] = \frac{E}{1-\nu^2} \begin{bmatrix} 1 & \nu & 0 \\ \nu & 1 & 0 \\ 0 & 0 & \frac{1-\nu}{2} \end{bmatrix} \quad (8.6c)$$

Now we place forces at each node to equilibrate the boundary loads and external loads on the element. We call these

$$\{F\}^e = \begin{Bmatrix} F_i \\ F_j \\ F_m \end{Bmatrix} \quad (8.7a)$$

and again for plane stress each of these nodal forces has two components, for example,

$$F_i = \begin{Bmatrix} U_i \\ V_i \end{Bmatrix} \tag{8.7b}$$

where U_i and V_i are the x and y components, respectively, of the nodal force at i, as shown in Fig. 8.1. There are also body loads defined by

$$p = \begin{Bmatrix} X \\ Y \end{Bmatrix} \tag{8.7c}$$

To make the nodal forces statically equivalent to the boundary loads and external loads we use the principle of virtual work. Thus we impose a virtual displacement on the element and then equate the internal and external virtual work. In terms of Eq. (8.1) the virtual displacement is $d\{\delta\}^e$. The virtual displacements and strains in the element due to the latter are, from Eqs. (8.4c) and (8.5c),

$$d\{\phi\} = [N] \, d\{\delta\}^e$$

$$d\{\varepsilon\} = [B] \, d\{\delta\}^e \tag{8.8a}$$

The virtual work done by the nodal forces is then

$$dW_F = (d\{\delta\}^e)^T \{F\}^e \tag{8.8b}$$

where the T denotes "the transpose of." The internal virtual work is, similarly

$$dW_I = (d\{\varepsilon\})^T \{\sigma\} - (d\{\phi\})^T \{p\} \tag{8.8c}$$

where the first term is the strain energy and the second term is the virtual work of the body forces. Now substitute for the strain from Eq. (8.5c) and the displacement from Eq. (8.4c) to obtain, after some re-arrangement,

$$dW_I = (d\{\delta\}^e)^T ([B]^T \{\sigma\} - [N]^T \{p\}) \tag{8.8d}$$

To apply the principle of virtual work we equate the internal and external work to give

$$(d\{\delta\}^e)^T \{F\}^e = (d\{\delta\}^e)^T \int_V ([B]^T \{\sigma\} - [N]^T \{p\}) \, dV \tag{8.8e}$$

The integral on the right signifies the total internal virtual work in the element. There is no integral on the left because the nodal forces are discrete values. For an arbitrary virtual displacement the two sides of Eq. (8.8e) must be equal, so the result is

$$\{F\}^e = \int_V [B]^T [D][B]\{\delta\}^e \, dV$$

$$- \int_V [N]^T \{p\} \, dV \tag{8.8f}$$

where we have used Eqs. (8.5c) and (8.6b). Since $\{\delta\}^e$ is a collection of numbers it doesn't take part in the integration, hence we define

$$[k]^e = \int_V [B]^T [D][B] \, dV \tag{8.8g}$$

This is called the stiffness matrix of the element. We also define

$$\{F\}_p^e = \int_V [N]^T \{p\} \, dV \tag{8.8h}$$

which is simply the matrix of nodal forces due to distributed body loads. With these definitions we may write

$$\{F\}^e = [k]^e \{\delta\}^e - \{F\}_p^e \tag{8.8i}$$

The foregoing development pertains to one element. Now we must assemble the elements to represent the total body under consideration. To do this we observe first that we may have concentrated forces applied at the nodes. Furthermore, at the boundaries of the body we may specify either the displacements or the distributed force per unit area. In the latter case elements at boundaries will have nodal loads caused by the external distributed loads. If the distributed force is $\{g\}$ per unit area we then would have to add the following term to the previous development

$$\{F\}_b^e = \int_A [N]^T \{g\} \, dA \tag{8.9a}$$

where the integration extends over the boundary area of an element at the boundary. This equation gives the forces at the nodes due to the distributed surface loads. With this Eq. (8.8i) becomes

$$\{F\}^e = [k]^e \{\delta\}^e - \{F\}_p^e - \{F\}_b^e \tag{8.9b}$$

At this point we can abandon the inter-element force idea and apply the previous development to the entire assembly of elements. Thus we write

$$\{\phi\} = \{\overline{N}\}\{\delta\} \qquad (8.10a)$$

where $\{\delta\}$ refers to the displacements at all of the nodal points and $\{\overline{N}\}$ refers to $\overline{N}_i = N_i^e$. Here e identifies the element and i identifies the point. If i is not in e, $N_i^e = 0 = \overline{N}_i$. In a similar fashion

$$\{\varepsilon\} = [\overline{B}]\{\delta\} \qquad (8.10b)$$

Now, the virtual work principle gives

$$(d\{\delta\})^T\{R\} + \int_V (d\{\phi\})^T\{p\}\, dV$$

$$+ \int_A (d\{\phi\})^T\{g\}\, dA = \int_V (d\{\varepsilon\})^T\{\sigma\}\, dV \qquad (8.10c)$$

where $\{R\}$ is the vector of all of the force components at all of the nodes. With Eqs. (8.10a, b) the foregoing equation eventually leads to

$$\{R\} = [K]\{\delta\} - \{F\}_p - \{F\}_b \qquad (8.10d)$$

where all quantities are obtained from the element quantities by summation, for example,

$$[K] = \sum [k]^e \qquad (8.10e)$$

The unknown nodal displacements can now be obtained from Eq. (8.10d) by standard techniques involving matrix algebra. For example, Eq. (8.10d) can be partitioned so that

$$\left\{\begin{array}{c} P \\ \hline Q \end{array}\right\} = \left(\begin{array}{c|c} K_{11} & K_{12} \\ \hline K_{21} & K_{22} \end{array}\right) \left\{\begin{array}{c} w \\ \hline C \end{array}\right\} - \left\{\begin{array}{c} F_a \\ \hline F_b \end{array}\right\}_p - \left\{\begin{array}{c} F_a \\ \hline F_b \end{array}\right\}_b \qquad (8.11a)$$

where $\{P\}$ represents any loads that are applied to the nodes, $\{Q\}$ are the unknown reactions where displacements are specified, $\{w\}$ are the unknown nodal displacements, $\{C\}$ are the specified nodal displacements, if any, $\{F_a\}$ are the nodal distributed and body forces for nodes with applied loads, and $\{F_b\}$ are the nodal distributed and body forces for nodes with applied displacements. Eqs. (8.11a) can be solved for the unknown reactions and displacements to give

$$\{Q\} = [K_{21}]\{w\} + [K_{22}]\{C\} - \{F_b\}_p - \{F_b\}_b$$

$$\{w\} = [K_{11}]^{-1}\{\{P\} + \{F_a\}_p + \{F_a\}_b - [K_{12}]\{C\}\} \qquad (8.11b)$$

Finally, the stresses in the assembly are found from

$$\{\sigma\} = [D][\bar{B}]\{\delta\} = [S]\{\delta\} \qquad (8.12)$$

where $[S]$ is called the stress matrix of the structure.

This completes our brief summary of the finite element method. While it has been restricted to the use of the constant strain triangle for plane stress problems, the development can easily be extended to other classes of problems and elements. The extension to these can be effected by modification of Eq. (8.2), and relationships that derive from it, and by alteration of the $[\bar{B}]$ and $[\bar{D}]$ matrices. The manipulations are, however, identical. For further details the reader is urged to consult Zienkiewicz [1], Gallagher [2], Ural [3], and other books. In the next section we consider the incorporation of creep into the calculations.

8.3 APPLICATION TO CREEP ANALYSIS

The previous presentation of the finite element method pertained to elastic behavior. It can be readily extended to include creep and thermal effects by noting that the total strain can be decomposed as follows (see Chapter 2, Eq. (2.38) with plastic strains not considered):

$$\{\varepsilon\} = \{\varepsilon^E\} + \{\varepsilon^C\} + \{\alpha T\} \qquad (8.13a)$$

from which the elastic strains are obtained as

$$\{\varepsilon^E\} = \{\varepsilon\} - \{\varepsilon^C\} - \{\alpha T\} \qquad (8.13b)$$

If we note that the development in Section 8.2 governs the elastic strains, we may substitute Eq. (8.13b) where appropriate to obtain

$$\{\sigma\} = [\bar{D}](\{\varepsilon\} - \{\varepsilon^C\} - \{\alpha T\})$$

$$\{R\} = [K]\{\delta\} - \{F\}_p - \{F\}_b - [B]^T[D](\{\varepsilon^C\} + \{\alpha T\}) \qquad (8.13c)$$

in which $\{\varepsilon\}$ is the total strain. It is this quantity that is always related to the displacements through equations such as Eqs. (8.5a, b, c) and (8.10b).

8.3.1 Method of Initial Strains

Examination of the first of Eq. (8.13c) reveals that the presence of $\{\varepsilon^C\}$ is similar to an initial strain that is subtracted from the total strain $\{\varepsilon\}$. Alternatively, $\{\varepsilon^C\}$ is like the thermal strain $\{\alpha T\}$ that also appears. For these reasons the method that is about to be described has been called by various authors the method of initial strains or the method of thermal strains. We adopt the former usage because it predominates in the current literature. The method was proposed first by Mendelson, Hirschberg, and Manson [4] and proceeds in a stepwise fashion as follows:

Step 1 At time $t=0$ determine the elastic solution of the problem by setting $\{\varepsilon^C\}=0$ in Eqs. (8.13c). This gives the nodal displacements and stresses at the start of the problem.

Step 2 Assume that the stresses obtained in Step 1 remain constant over a small increment of time Δt and calculate the increments of creep strain that occur in the increment of time by using Eqs. (2.40) that are repeated here in the form

$$\Delta \varepsilon_{ij}^C = \frac{3}{2} \frac{d\bar{\varepsilon}^C}{dt} \frac{S_{ij}}{\sigma_e} \Delta t$$

$$\bar{\varepsilon}^C = \bar{\varepsilon}^C(\sigma_e, T, t) \qquad (8.14)$$

Step 3 Use the creep strain increments obtained in Step 2 to determine the new creep strains at the end of the time increment by means of

$$\varepsilon_{ij}^C(t+\Delta t) = \varepsilon_{ij}^C(t) + \Delta \varepsilon_{ij}^C \qquad (8.14a)$$

Then determine the stresses at the end of the time increment by substituting the creep strains into Eqs. (8.13c) and solving them.

Step 4 Test the stresses against their existing values at the beginning of the time increment. If they are larger than a preset fraction of the existing stresses, repeat Steps 2 and 3 with a smaller time increment. If the stresses are less than or equal to the preset fraction of the existing stresses, go to Step 5.

Step 5 Add another time increment and repeat Steps 2, 3, 4. Continue in this manner until the desired time interval has been reached or until a stationary state has been achieved.

It is clear from Steps 1 and 3 that the above method has as its foundation an elastic analysis. Thus we see that creep programs represent extensions of existing elastic programs. If the loads and temperatures are also changing with time, this can be incorporated into the analysis by using their appropriate values at the time of interest in Step 3. Moreover, it can be seen from Eqs. (8.11b) and (8.13c) that any matrix inversions that are required need to be done only once, since the stiffness matrix does not change from time increment to time increment.

The method of initial strains, as just described, has been widely accepted and is used in all contemporary computer programs for creep analysis. Its only drawback is in its convergence to a solution. This is manifested in Step 4, where the stresses are tested against their existing values. The necessity that they be less than a preset fraction of the existing stresses at the beginning of a time increment makes the solution subject to a great deal of judgment. For example, what should the fraction be? In setting it one may lose accuracy if it is "too large", and one may add greatly to the running time if it is "too small". The matter of losing accuracy is also intriguing, since one doesn't know the exact solution in advance. Then how is one to know that accuracy is being lost? The only possibility for avoiding such a difficulty is to run the problem with several preset fractions to see if, after a certain point, the answers do not change with further tightening of the tolerance. It is in this area where current effort in creep analysis by numerical methods is being exerted. That is, can a reliable method for choosing the time increment, for example, be found so that the calculation can be carried out with confidence? The answer has not yet been resolved, although several possibilities have been proposed.

In one method a weighted average of the initial strains from previous iterations within Step 4 of the procedure is used. Other methods involve built-in alteration of the time increment when the ratio of the effective creep strain increment to the initial effective elastic strain is greater than a preset amount.

8.3.2 Extraction of Time for Strain Hardening Creep

In Chapter 2 we mentioned that the strain hardening assumption seems to be the most practical and realistic approach to variable stress creep problems at the present time. We showed that if the basic creep equation, Eq. (2.3), say, is solved for time and the result is substituted into the rate form of the creep equation, Eq. (2.4), we obtain the strain hardening creep equation, Eq. (2.6), in one dimension. We also showed that in many cases more complicated creep equations such as Eq. (2.7) are assumed, that is,

$$\varepsilon^C = \sum_{i=1}^{N} A_i(\sigma)(1 - e^{-r_i(\sigma)t}) + \dot{\varepsilon}_m t \qquad (8.15)$$

This particular form has been recommended for use with 304 and 316 stainless steels by the Oak Ridge National Laboratory [5]. At the present time N is usually taken as 2, for example, Eq. (2.9).

If an expression such as Eq. (8.15) or any other expression from which time cannot be readily extracted is used, a numerical procedure must be employed. The method chosen is usually Newton's method [16]. This is a procedure for finding roots of equations. Consider a function $f(t)$ and a point t_0 which is not a root of $f(t)$ but is "reasonably close" to a root. Expand $f(t)$ in a Taylor series about t_0, where a prime denotes a derivative with respect to t:

$$f(t) = f(t_0) + (t - t_0) f'(t_0) + \frac{(t - t_0)^2}{2!} f''(t_0) + \ldots \qquad (8.16)$$

If $f(t)$ is set to zero then t must be a root and the right-hand side of the expansion constitutes an equation for the root t. Unfortunately, the equation is a polynomial of infinite order. However, an approximate value of the root t can be obtained by setting $f(t)$ to zero and taking only the first two (linear) terms on the right hand side to give

$$f(t_0) + (t - t_0) f'(t_0) = 0 \qquad (8.16a)$$

This may be solved for t to give

$$t = t_0 - \frac{f(t_0)}{f'(t_0)} \qquad (8.16b)$$

Now t is an improvement over t_0. We can thus replace t_0 in Eq. (8.16b) with it and obtain another estimate. The $(n+1)$st estimate can thus be written in terms of the nth estimate as

$$t_{n+1} = t_n - \frac{f(t_n)}{f'(t_n)} \qquad (8.16c)$$

The convergence is usually rapid. In our case the recursion formula for a creep law such as Eq. (8.15) would be

$$t_{n+1}=t_n+\frac{\varepsilon^C-\sum_{i=1}^{N}A_i(1-e^{-r_it_n})-\dot{\varepsilon}_m t_n}{\sum_{i=1}^{N}A_i r_i e^{-r_it_n}+\dot{\varepsilon}_m} \qquad (8.16d)$$

Then, after t has converged it is substituted into the rate form of Eq. (8.15)

$$\dot{\varepsilon}^C=\sum_{i=1}^{N}A_i r_i e^{-r_it}+\dot{\varepsilon}_m \qquad (8.16e)$$

Since the creep problem is being solved numerically to start with, the additional numerical analysis that is introduced by extracting the time with Newton's method presents no serious difficulties.

In writing Eq. (8.16e) we have a choice of selecting the time for the strain hardening model. As written here, the time solution was obtained from the *total* creep strain in Eq. (8.15). It is also possible to base the strain hardening on the transient creep strain only. In this case Eq. (8.16d) would be replaced by

$$t_{n+1}=t_n+\frac{\varepsilon^C-\sum_{i=1}^{N}A_i(1-e^{-r_it_n})}{\sum_{i=1}^{N}A_i r_i e^{-r_it_n}} \qquad (8.16f)$$

where we have deleted the secondary strain $\dot{\varepsilon}_m t$. Equation (8.16e) would remain the same, but the latter value of time would be used. Studies by Pugh [5] seem to indicate that there is not much difference in the predictions obtained by these two approaches for variable stress tests that were conducted at the Oak Ridge National Laboratory.

8.3.3 Computer Programs for Creep Analysis

We must emphasize that, as presented so far, the discussion is limited to elasticity and creep for small strains and displacements. This choice was made because it permitted us to concentrate on the creep problem rather than on the computing problem. The restriction is not necessary since programs are available that are capable of treating combined elastic-plastic-creep problems with large strain and displacements. The basis of one such program has been described by Haisler and Sanders [7]. Zienkiewicz's text [1] also discusses the necessary extensions for plasticity and large strains and displacements.

At this time, it is well to discuss some of the available general purpose computer programs that have the capability of performing creep analysis. Many such programs are in use and have been previously summarized by Nickell [8], Dhalla and Gallagher [9], and Yamada and Nagato [10]. We cite those most widely used today. The programs and their originators are listed below:

UNITED STATES

ADINA	K. J. Bathe, Massachusetts Institute of Technology, Cambridge, MA 02139
ANSYS	Swanson Analysis Systems, Inc., Houston, PA 15342
MARC	MARC Analysis Research Corp., Palo Alto, CA 95306
NEPSAP	D. N. Yates, Lockheed Missiles and Space Co., Sunnyvale, CA 94088
WECAN	Anon., Westinghouse Research and Development Center, Pittsburgh, PA 15235

UNITED KINGDOM

ASAS	R. K. Henrywood, Atkins Research and Development, Epsom, Surrey
BERSAFE	T. K. Hellen, Central Electricity Generating Board, Berkeley, Gloucestershire
FESS	O. C. Zienkiewicz, University of Wales, Swansea
PAFEC	R. D. Henshell, University of Nottingham, Nottingham

GERMANY

ASKA	J. H. Argyris, Institut für Statik und Dynamik der Luft-und-Raumfahrtkonstruktionen, Stuttgart

JAPAN

ASTUC	Y. Yokouchi, A. Ishii and T. Edamoto, University of Electro-Communications
MINAT-X	K. Nagato and A. Minato, Kawasaki Heavy Industries
PCRAP-2	T. Mori and T. Murakami, Toshiba Corp.
SATEPIC	M. Ueda and M. Tanikawa, Hitachi Shipbuilding and Engineering Co., Ltd.

TABLE 8.1 Comparison of Programs Used in the United States

Capability	Program					
	ADINA	ANSYS	MARC	NEPSAP	PLANS	WECAN
Type of Analysis						
Small deformation (inelastic)	X	X	X	X	X	X
Creep buckling	X	X	X	X	0	0
Large strain	X	0	0	X	X	0
Heat transfer	X	X	X	0	0	X
Material Models						
Plasticity						
Isotropic hardening	X	X	X	X	0	X
Kinematic hardening	X	X	X	X	X	X
Creep						
Equation of state (strain hardening)	X	X	X	X	X	X
Auxiliary rules for stress reversal	X	X	X	0	X	X
User supplied creep equations	0	0	X	0	X	X
Temperature dependent properties	X	X	X	X	0	X
Orthotropic properties	X	X	X	X	X	X

Now let us discuss some of the pertinent capabilities of the *American* programs as surveyed by Dhalla and Gallagher [9]. We will only consider several categories that are concerned with creep as listed below. In Table 8.1 an X indicates that the program has a given capability while a 0 indicates that it does not.

As far as creep is concerned, all of these use the initial strain method except ADINA. All but WECAN have means of setting the convergence of the creep analysis. Obviously, the programs have extensive capabilities in the analysis of the types of problems that might arise in creep analysis of solids and structures. They also have extensive element libraries and the ability to carry out dynamic analysis. The main difficulty with the general purpose finite element programs is that they are usually proprietary. Thus they are not readily accessible to the user who cannot afford to pay for them. To such users they remain "black boxes." To large corporate users they become expensive analytical tools that many have purchased.

Up to this point, our entire discussion of creep analysis by the digital computer has been slanted toward the use of general purpose finite element programs. It is not always necessary to use such a powerful tool. Sometimes, as we will illustrate, a so-called special purpose program for a given problem can be written. In such a case the program solves equations that pertain only to the problem at hand. The initial strain method is, however, still used to handle the creep. Only the equations being manipulated are different. The benefit of a special purpose program is that it uses less computer time to solve a given problem than if that problem were solved with a general purpose program. Special purpose programs can also be viewed as approximate solutions, in which some type of structural idealization has been made to reduce the complexity of analysis. Such programs provide accurate solutions for problems that fit the structural idealization, and approximate solutions for problems that do not. In the next section we present the results of some numerical analyses of creep problems.

8.4 ILLUSTRATIVE SOLUTIONS

In the foregoing discussion we have presented the development of numerical solutions of creep problems along the lines of general purpose finite element programs. At this point it is well to demonstrate the type of problems that can be solved today with both general purpose and special purpose programs. Wherever possible we shall, in doing this, also present experimental data that pertain to the problems. The examples to be presented are (a) Stress redistribution in a thick pressure vessel with an ellipsoidal head, (b) Stress redistribution in a rotating disc, (c) Creep ratchetting of a thick tube, (d) Creep of a straight tube under combined bending, pressure and thermal loads, (e) Creep and relaxation of a circular plate under prescribed transverse deflections, (f) Creep buckling of an axially compressed cylindrical shell, (g) Creep of a rotating gas turbine seal ring and (h) Cyclic creep of a primary closure seal for a nuclear power plant component.

8.4.1 Stress Redistribution in a Thick Pressure Vessel with an Ellipsoidal Head

One of the earliest creep analyses to be performed with the digital computer was an analysis of a thick pressure vessel with an ellipsoidal head. The analysis was done by Greenbaum and Rubinstein [11]. In their calculations the finite element method in conjunction with a time hardening creep law was used and the stress and creep strain histories were determined for a constant pressure loading. We have reproduced, in Figs. 8.2 to 8.4, some of the results that were obtained. Of these, Fig. 8.2 shows

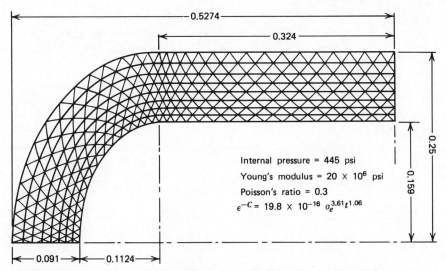

Fig. 8.2 Geometry, properties, and finite element breakdown of a pressure vessel with an ellipsoidal head [11]. Courtesy of the American Society of Mechanical Engineers.

the dimensions of the vessel and head, the mechanical properties, and the breakdown of the assembly into a collection of 520 triangular ring elements. Then, in Fig. 8.3 and 8.4 the effective stress contours at $t=0$ (the elastic solution) and $t=3$ hr. (the creep solution) are plotted. In Fig. 8.5 the effective stress histories at the junctions of cylindrical vessels with spherical, ellipsoidal, and flat heads are compared. It is seen that the stationary

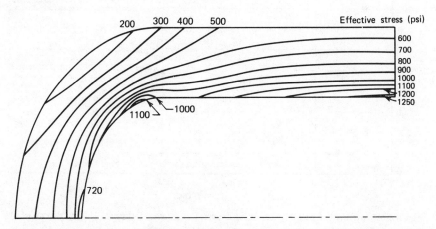

Fig. 8.3 Effective stress contours at $t=0$ for the vessel of Fig. 8.2 [11]. Courtesy of the American Society of Mechanical Engineers.

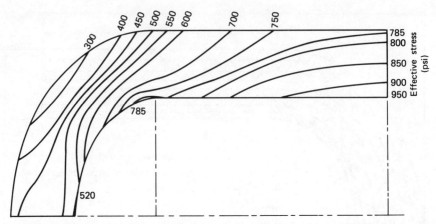

Fig. 8.4 Effective stress contours at $t=3$ hours for the vessel of Fig. 8.2 [11]. Courtesy of the American Society of Mechanical Engineers.

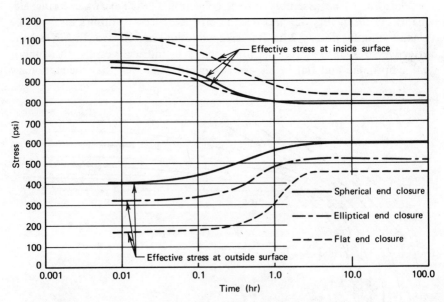

Fig. 8.5 Stress history at the junction of a vessel and several closures [11]. Courtesy of the American Society of Mechanical Engineers.

state was reached in about three hours for each head. Fig. 8.5 also shows a common result of stress redistribution studies, namely, that the highest and the lowest stresses usually undergo the most change during redistribution. It is noted that time hardening was used in the calculation. Since the load was steady, this is not felt to have too severe an effect upon the results.

8.4.2 Stress Redistribution in a Rotating Disc

Wahl and his colleagues carried out creep tests of rotating discs in 1954 [12]. Later, the tests were analyzed by Mendelson et al. [4] in their original presentation of the method of initial strains. Their solution was based on what would be called a special purpose program in today's usage. The first basic equation of the problem of a rotating thin disc in plane stress is the equilibrium equation

$$\frac{d}{dr}(hr\sigma_r) - h\sigma_\theta + \rho\omega^2 r^2 h = 0 \tag{8.17}$$

where r and h are the radius and thickness, σ_r and σ_θ are the radial and tangential stresses, and ρ and ω are the density and angular velocity. The second basic equation is the compatibility equation

$$\varepsilon_r - \varepsilon_\theta = r\frac{d\varepsilon_\theta}{dr} \tag{8.17a}$$

where ε_r and ε_θ are the radial and tangential strains. The stress-strain relations are

$$\varepsilon_r = \frac{1}{E}(\sigma_r - \nu\sigma_\theta) + \alpha T + \varepsilon_r^C + \Delta\varepsilon_r^C$$

$$\varepsilon_\theta = \frac{1}{E}(\sigma_\theta - \nu\sigma_r) + \alpha T + \varepsilon_\theta^C + \Delta\varepsilon_\theta^C \tag{8.17b}$$

Here ε_r^C and ε_θ^C are the accumulated creep strains, and $\Delta\varepsilon_r^C$ and $\Delta\varepsilon_\theta^C$ are the creep strain increments in the current time increment. The creep strain increments are related to the stresses by

$$\Delta\varepsilon_r^C = \frac{\Delta\bar{\varepsilon}^C}{2\sigma_e}(2\sigma_r - \sigma_\theta)$$

$$\Delta\varepsilon_\theta^C = \frac{\Delta\bar{\varepsilon}^C}{2\sigma_e}(2\sigma_\theta - \sigma_r)$$

$$\Delta\varepsilon_z^C = -\Delta\varepsilon_r^C - \Delta\varepsilon_\theta^C \tag{8.17c}$$

Numerical Analysis of Creep

The last of these expresses the constant volume condition, while $\Delta \bar{\varepsilon}^C$ and σ_e are the effective creep strain increment and stress.

Now, if we substitute Eqs. (8.17b) into Eqs. (8.17a) we obtain, in addition to the equilibrium equation, the second equation that governs the stresses

$$\frac{d}{dr}\left(\frac{\sigma_\theta}{E} - \frac{\nu\sigma_r}{E} + \alpha T + \varepsilon_\theta^C + \Delta\varepsilon_\theta^C\right) = \frac{1+\nu}{E}\frac{\sigma_r - \sigma_\theta}{r} + \frac{\varepsilon_r^C - \varepsilon_\theta^C}{r} + \frac{\Delta\varepsilon_r^C - \Delta\varepsilon_\theta^C}{r} \tag{8.17d}$$

Since the disc dimensions change with time, the values of r and h change from their original values R and H according to the formula

$$h = \frac{H}{(1+\varepsilon_\theta^C)(1+\varepsilon_r^C)}$$

$$r = R(1+\varepsilon_r^C) \tag{8.17e}$$

These are based on the definition of radial strain and the constancy of volume during flow in creep.

The experimental disc had an outer diameter of 12 in., an inner diameter of 2.5 in., and was rotated at 15,000 rpm. Young's modulus was $E = 18(10^6)$ psi at the temperature of 1000°F. The creep relationship was taken to be

$$\Delta\bar{\varepsilon}^C = 4.41(10^{-32})\sigma_e^{6.2}\Delta t \tag{8.17f}$$

Note that this is a steady creep law. It was chosen because the tests were long term at constant load. A second calculation was carried out with

$$\bar{\varepsilon}^C = A\sigma_e^m t^n \tag{8.17g}$$

where $A = 1.5(10^{-32})$, $m = 6$, $n = 2/3$. Both time hardening and strain hardening were used. To solve the equations of the problem, they were written in finite difference form and solved by the initial strain method. In the present context the steps were:

1. A time increment Δt was chosen, the accumulated creep strains ε_r^C and ε_θ^C were taken as zero, and initial guesses of $\Delta\varepsilon_r^C = \Delta\varepsilon_\theta^C = 10^{-5}$ were made at every radial position. $\Delta\bar{\varepsilon}^C$ was computed everywhere.
2. Eqs. (8.17) and (8.17d) were solved by the method of finite differences.

3. σ_e was computed from Eq. (8.17f) for stationary creep or Eq. (8.17g) for time hardening and strain hardening creep.
4. New approximations were obtained for $\Delta\varepsilon_r^C$, $\Delta\varepsilon_\theta^C$ from Eqs. (8.17c).
5. Steps 1 through 4 were repeated until the strain increments did not change.
6. Another time increment was added and Step 1 was repeated. This time, ε_r^C and ε_θ^C were no longer zero and the guesses for $\Delta\varepsilon_r^C$ and $\Delta\varepsilon_\theta^C$ were taken to be equal to those obtained in the previous interval.

The procedure was continued for 180 hours of elapsed time. The results are shown in Figs. 8.6 to 8.9. For the case of the steady creep law Eq. (8.17f), Fig. 8.6 gives the stress redistribution for the tangential stress at the bore and the rim of the disc, and Fig. 8.7 gives a comparison of the experimental and calculated strain distributions. In both of these figures the dotted lines are obtained with the elastic strain neglected and the solid lines are obtained with the elastic strains included. (Alternately, these are referred to as having transient effects neglected and included, respectively.) Then, Fig. 8.8 shows the change in the dimensions of the disc as time goes on for the same case of the steady creep law, Eq. (8.17f). In Fig. 8.6 it is seen that stationary creep is achieved in about 10 hours. Moreover, it shows that after a period of stationary creep the stresses begin rising. This is because, as the disc radius increases due to creep, as shown in Fig. 8.8, the centrifugal load increases and the thickness decreases. Thus the stress increases and eventually the disc fails. The comparison with the experimental results shown in Fig. 8.7 is fairly good. This, in conjunction with

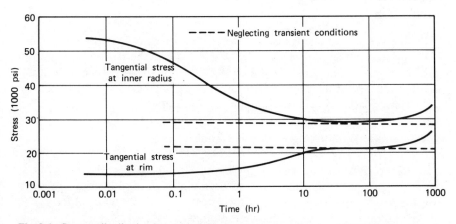

Fig. 8.6 Stress redistribution in rotating disc for steady creep law Eq. (8.17f) [4]. Courtesy of the American Society of Mechanical Engineers.

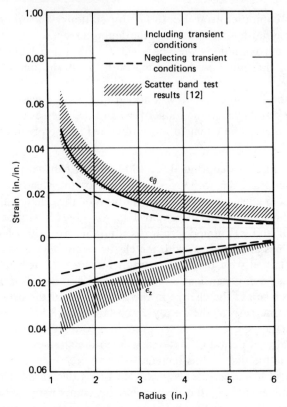

Fig. 8.7 Comparison of experimental and calculated strains at 180 hours for steady creep law Eq. (8.17f) [4]. Courtesy of the American Society of Mechanical Engineers.

Fig. 8.6, shows that the transient effects, that is, elastic strains, should not be neglected in the analysis. Finally, Fig. 8.9 shows the stress redistribution when the nonlinear creep law, Eq. (8.17g), is used with time and strain hardening. The difference in results is slight, owing to the fact that the centrifugal load is constant in the time period shown.

8.4.3 Creep Ratchetting of a Thick Tube

In Chapter 7 we discussed the problem of ratchetting with and without creep. To indicate that it has, in fact, been observed in industrial structures we described the results of an experiment of a thick tube under cyclic thermal loadings. These results were plotted in Fig. 7.11.

The pipe ratchetting test was carried out by Corum et al. [13], while the supporting numerical analyses were carried out by Clinard et al. [14]. The

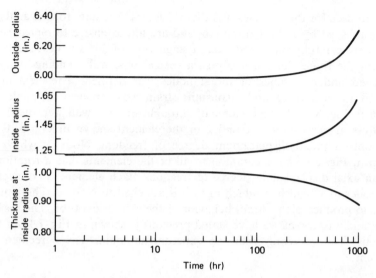

Fig. 8.8 Calculated variations in disc dimensions for steady creep law Eq. (8.17f) [4]. Courtesy of the American Society of Mechanical Engineers.

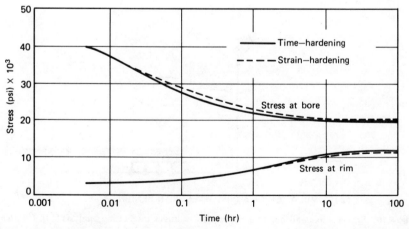

Fig. 8.9 Stress redistribution in rotating disc for nonlinear creep law Eq. (8.17g) [4]. Courtesy of the American Society of Mechanical Engineers.

program used for the analyses was PLACRE [15]. The data for the analysis were published by J. M. Corum [16] and are too extensive to repeat here. They were carefully measured from a single heat of 304 stainless steel. To perform the finite element analysis, a special pipe wall bending element was used and a number of radial nodes was defined. The element is referred to as a three node, constant strain axisymmetric rectangle, as shown in Fig. 8.10. Sixteen elements through the pipe wall were used for the stress analysis. Nodes 1 and 2 of the element shown in Fig. 8.10(a) have only a radial displacement degree of freedom. Node 3, which as shown in Fig. 8.10(b) is common to all of the elements, has a rotational and an axial displacement degree of freedom. Each element is connected by a roller to a rigid bar, which in turn is attached to Node 3. The aim of this is to produce plane strain behavior of the cylindrical tube. The results of the analysis are, as we have stated previously, shown in Fig. 7.11 of the previous chapter, along with the experimental data. The agreement is excellent.

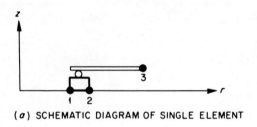

(a) SCHEMATIC DIAGRAM OF SINGLE ELEMENT

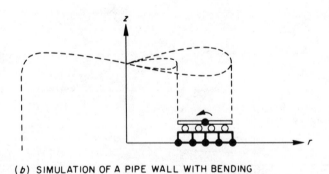

(b) SIMULATION OF A PIPE WALL WITH BENDING

Fig. 8.10 Special pipe wall bending element for cylinder ratchetting analysis [14]. Courtesy of the Oak Ridge National Laboratory.

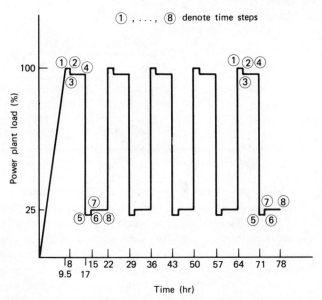

Fig. 8.11 Loading history for cylinder under combined loads [17]. Courtesy of the American Society of Mechanical Engineers.

8.4.4 Creep of a Straight Tube Under Combined Bending, Pressure and Thermal Loads

Chern and Pai [17] analyzed a reheat tube in the steam generator of an HTGR (High-Temperature Gas-Cooled Reactor). This tube is subjected to the loading shown in Fig. 8.11. At each numbered point in that figure, the mechanical loads and temperature undergo some change, as shown in Tables 8.2 and 8.3. Table 8.2 shows the mechanical loads, that is, internal and external pressure, axial compression, and applied curvature, while Table 8.3 shows the temperatures that exist.

Note that the first four cycles are repetitive, and the fifth cycle is different. The material of the tube is Incolloy 800 Grade 2.

TABLE 8.2 Mechanical Loads on Tube

Load	Internal Pressure (psi)	External Pressure (psi)	Axial Compression (lb)	Curvature (rad/in.)
Full	628	712	579	$7.99(10^{-4})$
Quarter	146	630	1098	$5.83(10^{-4})$

TABLE 8.3 Temperatures on Tube, degrees F

		\multicolumn{8}{c}{Time Step (see Fig. 8.11)}
Time	Surface	1	2	3	4	5	6	7	8
First 4 cycles	Outer	1241	1241	1216	1216	1103	1103	1128	1128
	Inner	1154	1154	1139	1139	1088	1088	1113	1113
Fifth cycle	Outer	1274	1274	1249	1249	1131	1131	1159	1159
	Inner	1172	1172	1157	1157	1109	1109	1131	1131

A special purpose program was written to analyze the bending of a thin tube [17]. The basic equations of the theory are the cylinder bending equations, with plasticity and creep included. The problem is set up to take advantage of the fact that all variables depend only on the circumferential coordinate and are independent of the axial coordinate. The details of the solution of these equations in the presence of plastic and creep strains are given by Chern and Pai [17]. They follow the standard methods of cylindrical shell theory, as described, for example, by Kraus [18], in conjunction with the initial strain method that has been described in the

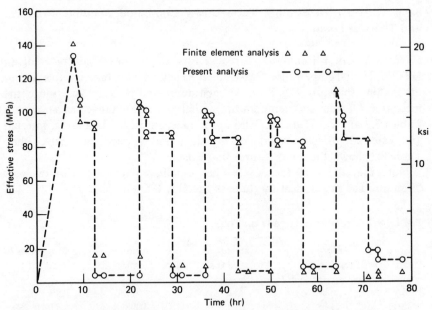

Fig. 8.12 Comparison of effective stresses at outside of cylinder under combined loads at critical point [17]. Courtesy of the American Society of Mechanical Engineers.

present chapter. The creep law is assumed to be of the form

$$\bar{\varepsilon}^C = A\sigma_e^\alpha t^\beta$$

along with the strain hardening hypothesis. The plasticity is handled by assuming a bilinear representation of the stress-strain curve with elastic slope E and the plastic slope E_p. Kinematic hardening is utilized.

The results of the five-cycle analysis corresponding to the load history of Fig. 8.11 and Tables 8.2 and 8.3 are shown in Fig. 8.12. There the effective stress at the outer surface of the tube is plotted against time. Two curves are shown. The one marked "present" refers to the solution just outlined. The one marked "finite element" refers to the results of the computation as obtained with a finite element program [19]. The agreement between the two calculations is very good for the most critical point in the tube. In assessing this, one should bear in mind that the tube was 0.138 in. thick and had a mean radius of 0.7435. Thus shell theory is not strictly applicable [18]. This explains why the present results and the finite element results are not exactly the same. The latter are based on a two-dimensional solid element.

8.4.5 Creep and Relaxation of a Circular Plate Under Prescribed Transverse Deflections

Corum and Richardson [20] conducted a series of experiments on beams and plates with both prescribed loads and displacements. As our next illustration of numerical analyses, we consider their tests of a simply supported circular plate to which a program of central deflection was applied. The circular plate specimen is shown in Fig. 8.14. The specimen was subjected to the deflection history shown in Fig. 8.13, where the peak

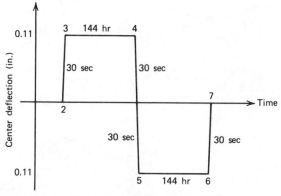

Fig. 8.13 Central deflection history for circular plate test [20]. Courtesy of the Oak Ridge National Laboratory.

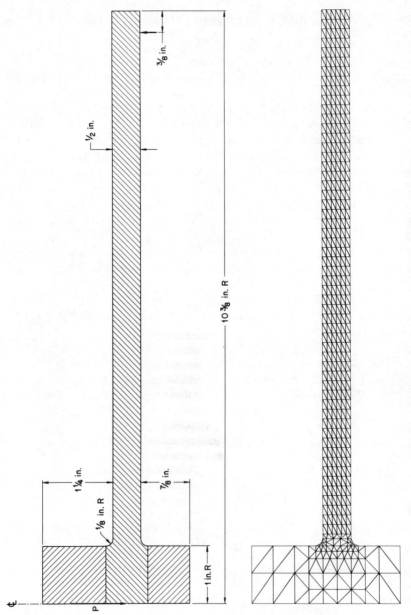

Fig. 8.14 Geometry and finite element breakup for circular plate test [14]. Courtesy of the Oak Ridge National Laboratory.

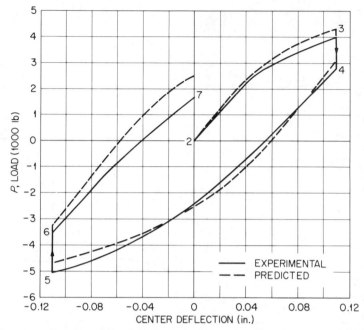

Fig. 8.15 Comparison of measured and predicted load-deflection response of circular plate. Numbered points correspond to points in the load history of Fig. 8.13 [14]. Courtesy of the Oak Ridge National Laboratory.

value of the center deflection was 0.11 in. The hold time periods shown in Fig. 8.14 were each nominally 144 hours and the plate deflection went from zero to maximum and back down in 30 sec. each.

The plate test was analyzed by Clinard et al. [14] using the finite element program, PLACRE [15]. The specimen and its finite element breakup are shown in Fig. 8.14. The model consisted of 784 three-node axisymmetric, triangular ring elements. Fig. 8.15 compares the calculated and the measured loads that were required to produce the short time central deflection changes. Fig. 8.16 compares the calculated and predicted load response during the two hold periods in the cycle. The agreement between the calculated and measured results is good. Fig. 8.16, in particular, demonstrates the adequacy of using the creep law for relaxation analysis, as discussed in Chapter 2.

8.4.6 Creep Buckling of an Axially Compressed Cylindrical Shell

In Chapter 6 we applied the methods that were described to the creep buckling tests of axially loaded shells carried out by Samuelson [21]. One

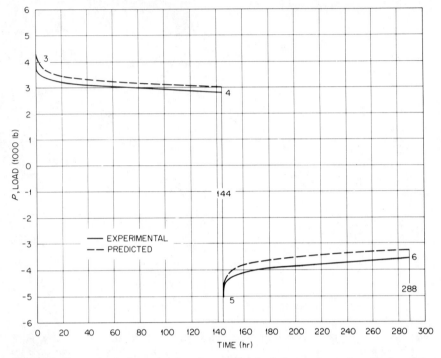

Fig. 8.16 Comparison of measured and predicted load-time response for circular plate. Numbered points correspond to points in the load history of Fig. 8.13 [14]. Courtesy of the Oak Ridge National Laboratory.

of these tests was analyzed by Stone and Nickell [22] with the finite element program, MARC, to which we referred in Section 8.3.2. In particular, they considered the case with radius to thickness ratio of 32 and applied axial stress equal to 12.1 kg/mm², which is included in Table 6.1 of Chapter 6. We may recall that the buckling time measured was 7.2 hours and there were several predictions based on the methods of Chapter 6 given as 11.5, 10.5, 9.12, and 8.30 hours.

In the finite element runs the creep was assumed to be stationary with a steady creep law

$$\dot{\varepsilon}^C = 4.4(10^{-10})\sigma_e^{5.8}$$

Two methods were used as follows, the details of which are beyond the scope of the text. The first is a step by step method:

1. Apply the external load to the shell and solve the time independent elasto-static problem.

2. Begin to march in time for creep analysis, as described earlier in this chapter.
3. Stop the creep analysis at some time, t.
4. Allow a small additional time increment for creep, Δt.
5. Solve the resulting eigenvalue problem for the eigenvalue λ of the load and the eigenfunction ϕ.
6. Calculate the critical time from the formula

$$t_{CR} = t + \lambda \, \Delta t$$

Stone and Nickell [22] applied this approach to a model that was made up of twenty elements along the length of the shell and three elements in the thickness direction. The assembly was analyzed for one hour of creep and the critical time was found to be 69.5 hours compared to the measured value of 7.2 hours. Next, the model was changed to 50 axial shell elements and again allowed to creep for one hour. The critical time was now reduced to 19.7 hours. This is still above the experimental value of 7.2 hours and all of the approximate values from Table 6.1.

Next, it was postulated that plasticity was occurring in the regions near the shell supports. The resulting behavior was analyzed by marching in time until the structural stiffness became non-positive definite. This gave a result of 9.56 hours, at which time a plastic hinge formed at one of the edges. The remaining differences, compared to the measured value of 7.2, were attributed to the presence of imperfections in the shell and inaccuracies in ignoring the transient creep behavior. Now, however, the finite element result is within the range of the approximate results obtained by the methods of Chapter 6.

It appears, therefore, that the approximate methods are to be preferred because of their relative simplicity and accuracy. As pointed out by Stone and Nickell [22], and by Dhalla and Gallagher [9], large deflection creep buckling analyses are currently still hazardous. Work is continuing in this area.

8.4.7 Creep of a Rotating Gas Turbine Seal Ring

In gas turbine construction there is a seal ring located in the vicinity of the point of attachment between the blades and the disc. The purpose of the seal ring is to contain cooling air while keeping out the hot gases that impinge on this general area. The assembly is shown in Fig. 8.17. As shown, the ring sits on a shoulder against which it is driven by the centrifugal force produced by rotation at 12,000 rpm. The highest temperature occurs at the outer tip of the ring. It is desired to maintain the shape of the ring so as to preserve the seal during operation of the turbine.

208 Numerical Analysis of Creep

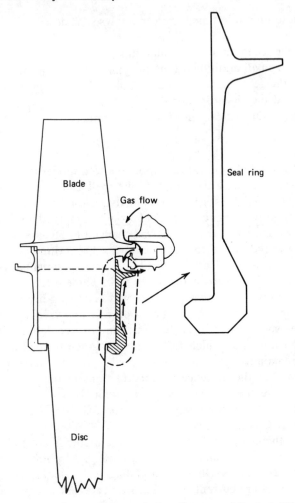

Fig. 8.17 Arrangement of gas turbine seal ring. Courtesy of Pratt & Whitney Aircraft Group.

The distortion of the seal ring was studied with the MARC general purpose finite element program that was listed in Section 8.3.2. To simplify matters only the hottest portions were modeled, as shown in Fig. 8.18. There three cases of temperature distribution ae described. The ring is made of a direct, hot, isostatically pressed nickel-base steel similar to IN 100. It was modeled as a collection of tetrahedral ring elements, as shown.

The creep strains at three interesting points for the three temperature distributions were calculated with MARC. The results for the 1365°F case

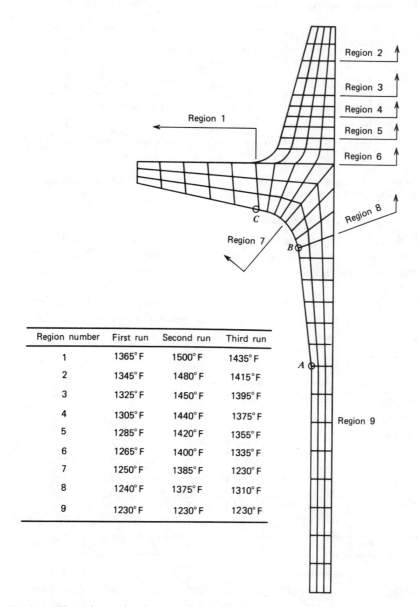

Fig. 8.18 Finite element breakup of seal ring. Courtesy of Pratt & Whitney Aircraft Group.

210   Numerical Analysis of Creep

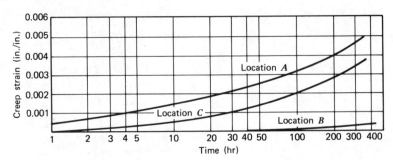

Fig. 8.19 Creep strain histories at critical locations for the 1365°F maximum temperature case. Courtesy of Pratt & Whitney Aircraft Group.

are plotted in Fig. 8.19, where the creep strain histories at the three locations are given. In another set of results the deflections of the tip of the "wing" were calculated for the three temperature cases. These are shown in Fig. 8.20.

We should note, in presenting this example, that it is in problems of irregular geometry such as this where the general purpose finite element programs demonstrate their greatest usefulness.

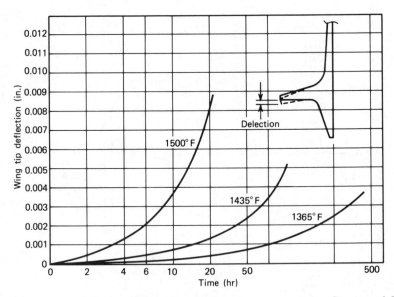

Fig. 8.20 Tip deflection histories for the three temperature cases. Courtesy of Pratt & Whitney Aircraft Group.

8.4.8 Cyclic Creep of a Primary Closure Seal for a Nuclear Power Plant Component

To achieve absolute sealing of the contents nuclear power plant components closures are welded with a seal such as the one shown in Fig. 8.21. In this case, taken from the FFTF (Fast Flux Test Facility) intermediate heat exchanger, a C-shaped closure is used. This toroidal element is welded to its adjacent members to provide the seal. In addition to resisting the pressure of the sodium coolant, the seal must be capable of accommodating differential movements between the structures to which it is attached.

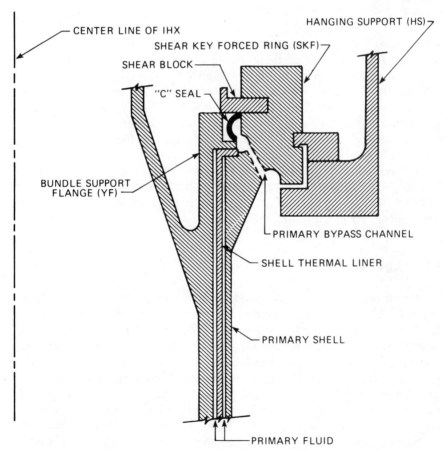

Fig. 8.21 Primary closure seal (C seal) support structure [24]. Courtesy of the Oak Ridge National Laboratory.

212 Numerical Analysis of Creep

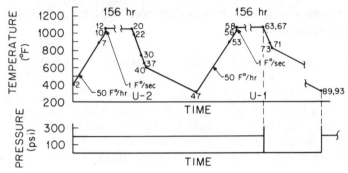

Fig. 8.22 Temperature and pressure histories for C-ring [24]. Courtesy of the Oak Ridge National Laboratory.

In this case both are massive and the ensuing analysis assumes that these members are not influenced by the seal. The C-seal is subjected to pressure and temperature histories of the type shown in Fig. 8.22. It was analyzed with the MARC finite element program by Dhalla [24], using the representation shown in Fig. 8.23. The stress-strain results for the meridional direction at the inside surface of Element 1 are shown in Fig. 8.24. In this case the loading was made up of a U2 cycle, three combined U2 and U1 cycles, and then three U1 cycles, all of which are defined in Fig. 8.22. The results shown in Fig. 8.24 indicate the complex nature of the response, including the fact that ratchetting is taking place in the C-seal.

This is another illustration of a case where the finite element approach demonstrates its greatest usefulness.

8.5 CLOSURE

In this chapter we have summarized the application of numerical analysis and the digital computer to the creep problem. As part of this we outlined the finite element method and the method of initial strains that is used to handle the creep aspect. We indicated some general purpose finite element programs that are in heavy use around the world today and that have creep analysis capabilities. Finally, we presented some problems that have been solved by the digital computer with either general purpose or special purpose programs. The presentation of these problems had several purposes. First, they showed what is possible in the way of solving complicated problems and loading histories. Second, they showed that they gave results that were reasonably close to available experimental results. Lastly, they were intended to provide benchmarks, that is, reference solutions for those who might have occasions to use a digital computer program on a problem

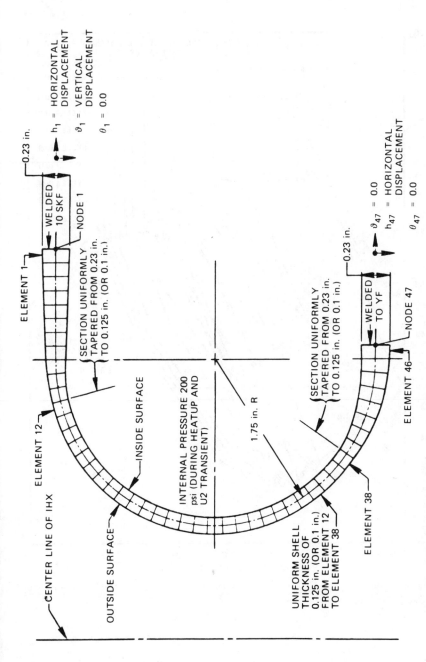

Fig. 8.23 Finite element idealization of C seal [24]. Courtesy of the Oak Ridge National Laboratory.

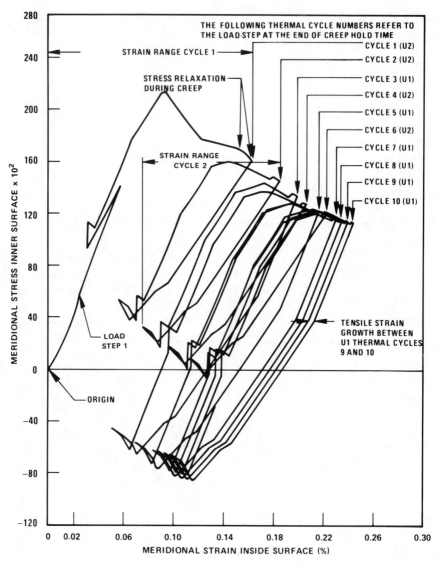

Fig. 8.24 Typical time-dependent stress-strain response curve for a C-ring [24]. Courtesy of the Oak Ridge National Laboratory.

of their own. In such an event it is always well to have a solution available that one can check some results against. We should add here that the booklet by Corum and Wright that contains references 13, 14, 15, 16, and 20 also contains the results of additional problems that one can utilize for this purpose. Recently, Kraus [23] compared the results of seven finite element analyses (two from the United States and five from Japan) and two approximate analyses (one each from the United States and the United Kingdom) to see how they compared to each other and to the results of the pipe ratchetting test by Corum et al [13]. The agreement was excellent and the comparison showed the value of the benchmark problem concept for verifying various analysis methods.

Many of the problems that we discussed included plastic behavior. This is the rule rather than the exception, and the student is well advised to become as familiar with this important aspect of the mechanical behavior of materials as with creep. Our presentation of plasticity in appropriate parts of Chapter 1, 2, and 7 provides a starting point for this. As we have stated several times, our preoccupation is limited to the creep aspect.

Although the numerical solutions have been shown to be accurate, they too have their limitations. Currently, the chief limitations are materials data, the proprietary nature of the general purpose computer programs, and the expense of digital computation. A subsidiary problem is the danger of indiscriminate use of programs by individuals who do not have a sufficiently thorough understanding of them. All of these areas are currently being studied. Better material models and further property data are actively being collected. The proprietary problem and the expense problem can possibly be handled by special purpose programs that are cheaper to run, subject to local control, but certainly not as versatile. The danger of indiscriminate use can be avoided by education of users via short courses put on by program developers.

In spite of the cited administrative drawbacks, it is clear the computer programs have become indispensable tools for engineers. They have made it possible to replace the drudgery of calculation with the more important and rewarding task of interpreting the results and coming up with a rational design.

REFERENCES

[1] O. C. Zienkiewicz, *The Finite Element Method*, McGraw-Hill Book Co. Ltd., 3rd ed., London (1977).

[2] R. H. Gallagher, *Finite Element Analysis Fundamentals*, Prentice-Hall, Inc., Englewood Cliffs, N. J. (1975).

[3] O. Ural, *Finite Element Method*, Intext Educational Publishers, New York (1973).

[4] A. Mendelson, M. H. Hirschberg, and S. S. Manson, A General Approach to the Practical Solution of Creep Problems, *Trans. A.S.M.E.*, vol. 81D, pp. 585–589 (1959).

[5] C. E. Pugh, Constitutive Equations for Creep Analysis of Liquid Moderated Fast Breeder Reactor Components, in S. Y. Zamrik, and R. I. Jetter, editors, *Advances in Design for Elevated Temperature Environment*, A.S.M.E., New York, pp. 1–16 (1975).

[6] F. B. Hildebrand, *Advanced Calculus for Applications*, Prentice-Hall, Inc., 2nd ed., Englewood Cliffs, N. J., pp. 368–370 (1976).

[7] W. E. Haisler and D. R. Sanders, Elastic-Plastic-Creep-Large Strain Analysis at Elevated Temperatures by the Finite Element Method, *Computers and Structures*, vol. 10, pp. 375–382 (1979).

[8] R. E. Nickell, Thermal Stress and Creep, in W. Pilkey, K. Saczalski, and H. Schaeffer, editors, *Structural Mechanics Computer Programs*, University Press of Virginia, Charlottesville, pp. 103–122 (1974).

[9] A. K. Dhalla and R. H. Gallagher, Computational Methods for Structural Analysis, A.S.M.E. (to appear).

[10] Y. Yamada and K. Nagato, First Report of Japanese Co-operation in International Benchmark Project of Subcommittee on Elevated Temperature Design of the U.S. Pressure Vessel Research Committee, March 1979.

[11] G. A. Greenbaum and M. F. Rubinstein, Creep Analysis of Axisymmetric Bodies Using Finite Elements, *Nucl. Eng. Des.*, vol. 7, pp. 379–397 (1968).

[12] A. M. Wahl, G. A. Sankey, M. J. Manjoine, and E. Shoemaker, Creep Tests of Rotating Discs at Elevated Temperature and Comparison to Theory, *J. Appl. Mech.*, vol. 21, pp. 225–235 (1954).

[13] J. M. Corum, H. C. Young, and A. G. Grindell, Thermal Ratchetting in Pipes Subjected to Intermittent Thermal Downshocks at Elevated Temperatures, in J. M. Corum and W. B. Wright, editors, *Pressure Vessels and Piping: Verification and Qualification of Inelastic Analysis Computer Programs*, A.S.M.E., New York, pp. 47–58 (1975).

[14] J. A. Clinard, J. M. Corum, and W. K. Sartory, Comparison of Typical Inelastic Analysis Predictions with Benchmark Problem Experimental Results, in J. M. Corum and W. B. Wright, *op. cit.*, pp. 79–98.

[15] W. K. Sartory, Finite Element Program Documentation, "High Temperature Structural Design Methods for LMFBR Components", ORNL-TM-3736, Oak Ridge National Laboratory, Oak Ridge, TN. March 1972.

[16] J. M. Corum, Material Property Data for Elastic-Plastic-Creep Analyses of Benchmark Problems, in J. M. Corum, and W. B. Wright, editors, *op. cit.*, pp. 99–109.

[17] J. M. Chern, and D. H. Pai, Inelastic Analysis of a Straight Tube Under Combined Bending Pressure and Thermal Loads, *A.S.M.E. J. Press. Vessel Tech.*, vol. 97J, pp. 155–162 (1975).

[18] H. Kraus, *Thin Elastic Shells*, John Wiley and Sons, New York (1967).

[19] F. K. Tzung, and C. Charman, A User's Manual for the Two Dimensional Finite Element Program for Thermal-Elastic-Plastic-Creep Analysis, Gulf-GA-A12753, Gulf General Atomic Company, November 1973.

[20] J. M. Corum, and M. Richardson, Elevated Temperature Tests of Simply Supported Beams and Circular Plates Subjected to Time Varying Loadings, in J. M. Corum, and W. B. Wright, editors, op. cit., pp. 13–25.

[21] L. A. Samuelson, Experimental Investigation of Creep Buckling of Cylindrical Shells Under Axial Compression and Bending, *A.S.M.E. J. Eng. for Industry*, vol. 90, 589–595 (1968).

[22] C. M. Stone, and R. E. Nickell, An Evaluation of Design Safety Factors for Time Dependent Buckling, SAND76-0121, NUREG-0179, Sandia Laboratories, Albuquerque, NM, February 1977.

[23] H. Kraus, International Benchmark Project on Simplified Methods for Elevated Temperature Design Analysis: Problem I—The Oak Ridge Pipe Ratchetting Experiment, Welding Research Council Bulletin, May 1980.

[24] A. K. Dhalla, Effect of Yield Strength Variation on the Inelastic Response of a C-Ring, A.S.M.E. Paper 75-PVP-31 (1975).

CHAPTER 9

CREEP-FATIGUE INTERACTION

9.1 INTRODUCTION

As implied by the title, this book is concerned with the *analysis* of the various phenomena that come under the topic of creep. Thus in the preceding chapters we discussed the analysis of creep deformation, stress redistribution, stress relaxation, rupture, buckling, and ratchetting, by various techniques. We did not concern ourselves with the results that were obtained in each case. That is to say, we did not ask such questions as how much creep deformation or ratchetting is too much, or is a certain buckling or rupture time satisfactory? Generally, the answers to these questions depend on the application and we touched on these briefly in Chapter 1, that is, we do not want dimensions to change by more than some amount that depends on available clearances, preservation of flow paths and the like. Also, we do not want components to buckle or rupture in a time that is less than their intended design life. In each case a safety factor on total deformation or ratchetting is selected, as is a safety factor on the time to buckling or rupture. These usually depend on practice in a given industry and company, and are based on experience, accuracy of analytical models, and the level of reliability that is required for the equipment.

There is, however, one area mentioned in Chapter 1 that has not been taken up as yet. This concerns the case of cyclic loadings that lead to cyclic stresses and strains such as those introduced in the first chapter, that

A.S.M.E. Code Procedure

is, continuous cycling, strain controlled cycling with relaxation of stress during hold periods, and stress controlled cycling with creep during hold periods. These were cited as three idealizations of typical situations that may, in fact, involve all three types. Our purpose in this last chapter is to discuss methods for evaluating the life of structures under cycling loadings at elevated temperature. This area is not yet fully understood; hence, this chapter is intended as an introduction to it. The problem is to determine the life of a structure that is undergoing a certain loading history, called a histogram, for which the stress and strain histories have been obtained by any of the methods that have been previously discussed here. Most likely this will have been accomplished by a computer program. It is desired to evaluate the total damage, which consists of the fatigue damage that occurs because the loading is cyclic and the creep rupture damage that occurs during any hold times in the cycle. In general, these effects occur simultaneously. Some people use the terminology "creep-fatigue interaction" to describe this phenomenon while others prefer the terminology "elevated temperature fatigue." We choose the former here. The reader can select his or her preference after this exposure to the problem.

In this chapter we discuss the evaluation of the creep-fatigue interaction as follows. First, we present a method that has been adopted by one industry faced with the analysis of creeping structures, namely, the pressure vessel industry, particularly the segment that is concerned with nuclear reactor vessel design. This is the method adopted by the American Society of Mechanical Engineers' (A.S.M.E.) Boiler and Pressure Vessel Code in its so-called Code Case N-47*. Using this as an arbitrary starting point, we then present summaries of two other methods that have been proposed: strain range partitioning and frequency separation. In each case our point of view remains that of the analyst. Thus we do not delve into the collection of creep-fatigue data, and so on. Our purpose is simply to give the reader a starting point. An excellent, detailed review of the entire field is given by Coffin, Manson, Carden, Severud, and Greenstreet [1].

9.2 A.S.M.E. CODE PROCEDURE

As our first method, we present the procedure that is described in A.S.M.E. Boiler and Pressure Vessel Code Case N-47 [2]. We present this method because it represents the result of the effort of one industry, namely, the pressure vessel industry, to provide design criteria for itself in the area of

---

*This has previously been designated by the A.S.M.E. as Code Case 1331 then as Code Case 1592.

220 Creep-Fatigue Interaction

creep-fatigue interaction, among others. It is meant to be a beginning, not the last word. In fact the Code is constantly developing. Hence, it is possible that one of the methods to be described later may eventually replace the one that will be described first.

The method of A.S.M.E. Code Case N-47 is based on the existence of an interaction diagram such as that shown in Fig. 9.1 for the material under consideration. The curves shown represent conservative representations of experimental data for three alloys. They are plots of failure due to combinations of fatigue damage caused by cycling and creep rupture damage caused by hold periods. The quantity

$$\sum_{i=1}^{n} \frac{t_i}{t_{Ri}}$$

is the total creep damage for n hold periods t_i, where t_{Ri} is the creep rupture time for the i-th stress acting alone. We have already encountered

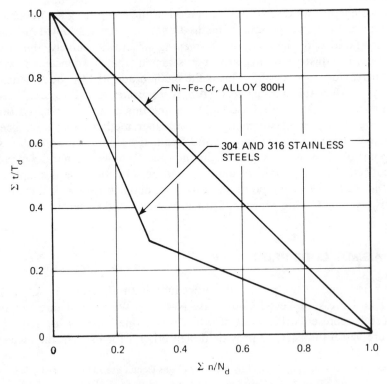

Fig. 9.1 Creep fatigue damage specified by ASME Pressure Vessel Code for use in linear damage creep fatigue analysis [2]. Courtesy of the Oak Ridge National Laboratory.

this quantity in Chapter 5, where we were concerned with creep rupture due to fluctuating stresses. Similarly, the quantity

$$\sum_{j=1}^{m} \frac{N_j}{N_{fj}}$$

is the total fatigue damage for m strain cycles N_j, where N_{fj} is the number of cycles to failure for the j-th cycle acting alone. m is not necessarily equal to n. The total damage must then be

$$\sum_{i=1}^{n} \frac{t_i}{t_{Ri}} + \sum_{j=1}^{m} \frac{N_j}{N_{fj}} \leqslant D \qquad (9.1)$$

where D is given in curves such as those shown in Fig. 9.1.

The abscissa in Fig. 9.1 pertains to fatigue damage without accompanying creep damage. The fatigue damage summation used here is known as Miner's rule [3]. Similarly, the ordinate in Fig. 9.1 pertains to rupture damage without accompanying fatigue damage. The rupture damage summation is known as Robinson's rule. When either type of damage, acting alone, reaches the value of 1.0 failure is said to occur. Combinations of fatigue and rupture damage that occur together are said to cause failure when their sum reaches the value D, which depends on the material. Thus combinations that fall below the curve for a given material are considered safe, whereas combinations that fall on or above the curve are considered to cause the structure to fail. The curve for alloy 800H in Fig. 9.1 is said to be a linear interaction curve, that is, the two types of damage are independent and when any combination adds up to one or more, failure is said to occur. The curve for 304 and 316 stainless steel is a non-linear interaction curve, that is, the creep damage and the fatigue damage interact, causing failure to occur for values of D that are less than one. In such a case the creep process deteriorates the fatigue process and vice versa, as opposed to the linear case where they do not influence each other.

Now let us outline the A.S.M.E. Code method. To do this we will assume that for some cyclic loading history we have at hand an accurate inelastically calculated stress and strain history. This may involve as many as six stress and six strain components and was presumably obtained for some component by use of a computer program such as the ones we discussed in Chapter 8*. The steps in the evaluation are as follows (see, e.g., [2, 17, and 18]):

---

*The Code Case method also permits an alternate procedure that is based on elastically calculated stress and strain histories.

Creep-Fatigue Interaction

Step 1. Calculate the history of the stress components σ_{ij} and the strain components ε_{ij} at every point of interest in the structure. This gives for example, a plot such as the one shown in Fig. 9.2 for *one cycle of one strain component*. Several cycles are involved if, for example, one is analyzing a component that is to be subjected to a normal operation cycle, an emergency operation cycle, an off peak operation cycle, and so on.

Step 2. Evaluate the fatigue damage:

(a) Designate $\varepsilon_{ij}^{(k)}$ as the strain associated with the k-th event in the cycle. In Fig. 9.2, k ranges from 1 to 17.
(b) Find the maximum value of $\varepsilon_{ij}^{(k)}$ and call this $\varepsilon_{ij}^{(l)}$. In Fig. 9.2 this would be $\varepsilon_{ij}^{(12)}$.
(c) Determine the strain range for each event by subtracting

$$\Delta\varepsilon_{ij}^{(k)} = \varepsilon_{ij}^{(l)} - \varepsilon_{ij}^{(k)}$$

(d) Repeat this for every strain component at every point in the structure for every event in the history.
(e) Calculate the equivalent strain range at every point for every event using the formula from Chapter 2

$$\Delta\bar{\varepsilon} = \frac{\sqrt{2}}{3}\left[(\Delta\varepsilon_{11}-\Delta\varepsilon_{22})^2 + (\Delta\varepsilon_{22}-\Delta\varepsilon_{33})^2 + (\Delta\varepsilon_{33}-\Delta\varepsilon_{11})^2 + 6(\Delta\varepsilon_{12}^2 + \Delta\varepsilon_{23}^2 + \Delta\varepsilon_{13}^2)\right]^{1/2}$$

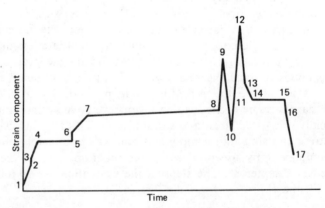

Fig. 9.2 Strain component histogram from inelastic analysis [18]. Courtesy of the American Society of Mechanical Engineers.

(f) Determine the maximum value of $\Delta\bar{\varepsilon}$ for the cycle at every point in the structure.
(g) Find the allowable number of cycles N_{fj} of this equivalent strain range at every point in the structure from experimental data such as those given in Fig. 9.3.
(h) Repeat the procedure for other cycles, if any.
(i) Calculate

$$\sum_{j=1}^{m} \frac{N_j}{N_{fj}}$$

for every point in the structure for all cycles. N_j is the number of times each cycle occurs.

Step 3. Evaluate the creep rupture damage:

(a) Calculate the effective stress at each point of the structure for the entire cycle using the formula from Chapter 2

$$\sigma_e = \frac{1}{\sqrt{2}} \left[(\sigma_{11} - \sigma_{22})^2 + (\sigma_{22} - \sigma_{33})^2 + (\sigma_{33} - \sigma_{11})^2 + 6(\sigma_{12}^2 + \sigma_{23}^2 + \sigma_{13}^2) \right]^{1/2}$$

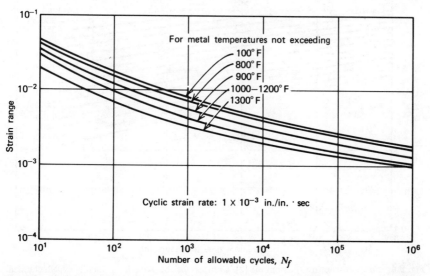

Fig. 9.3 Design fatigue strain range for 304 SS and 316 SS [2]. Courtesy of the American Society of Mechanical Engineers.

(b) Calculate the creep rupture stress at every point of the structure for the entire cycle from

$$S_{CR} = \frac{\sigma_e}{K}$$

Here K is a safety factor that is taken as 0.9 in the current version of the method [2].

(c) For each point in the structure and the entire cycle, determine the time to rupture t_{Ri} by entering a curve, such as one shown in Fig. 9.4, with S_{CR}.

(d) For each point in the structure and each time period, t_i in the cycle, calculate the creep rupture damage and sum the damage fractions for all of the cycles using

$$\sum_{i=1}^{n} \frac{t_i}{t_{Ri}}$$

Step 4. Evaluate the total damage at every point in the structure by entering a curve such as one in Fig. 9.1 for the material in question, using the creep and fatigue damage summations calculated before. If all

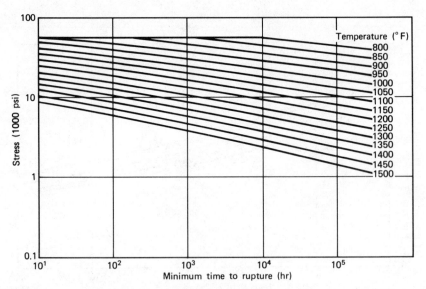

Fig. 9.4 Stress to rupture (minimum) [2]. Courtesy of the American Society of Mechanical Engineers.

of the plotted points fall below the damage curve, the structure is safe. If any plotted point falls on or above the curve, the structure is not safe.

It is obvious from the foregoing that there would be a great deal of data manipulation involved for a typically complex structure and general loading history. Usually these operations are programmed as post-processors in conjunction with computer programs such as those described in Chapter 8.

In reviewing the A.S.M.E. Code procedure several assumptions should be emphasized. First, as far as the stress and strain histories are concerned, strain range and time are the main variables in the damage evaluation. No cognizance is taken of the shape of the curves, the rates of change on the curves, and the order of occurence of the various events that cause fatigue damage and creep rupture damage. For various materials, as discussed by Coffin et al. [1], these factors have varying orders of importance. Thus in the sections that follow methods will be presented in which these factors are considered in the evaluation. The A.S.M.E. Code method tries to accommodate these factors by making damage curves such as one shown in Fig. 9.1 extremely conservative. Second, to determine the allowable number of cycles and the time to rupture for the various events it was assumed that effective values of the strain range and the stress govern the damage. This is to permit multiaxial stress states to be evaluated by data obtained from uniaxial tests. We have used this assumption on several occasions in this text. Zamrik [4] has shown, by running some biaxial fatigue tests, that it is fairly reasonable here as well.

Third, the method does not distinguish between compressive and tensile stresses. For some materials tensile stresses during hold times cause damage, while compressive stresses do not. For such materials the method is not likely to be accurate, although the results will be conservative. For other materials compressive stresses during hold times are damaging. In these cases the method should be realistic. Still other materials display "healing" due to compressive stresses. The method will be very conservative for these.

Fourth, it is assumed that the mean strain is constant, that is, there is no ratchetting. Ratchetting is handled by the methods of Chapter 7.

Kitagawa and Weeks [5] examined the procedure that has just been described for the case of hold time tests on 304 stainless steel. Their results are shown in Fig. 9.5, where calculated life is plotted against measured life. These show that the method is very conservative in that the measured lives are greater than the calculated lives.

To end the discussion, we may say that this method, although it is subject to certain assumptions, leads to very conservative predictions of life. It does this by a relatively simple procedure that is applied to available

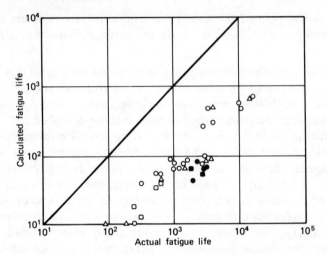

Fig. 9.5 Analysis of hold time data by time and cycle fraction rule according to ASME Code Case [5].

stress and strain histories. Moreover, most of the data that are required for its use are relatively easy to obtain. They are based on conventional fatigue and creep rupture tests. The damage interaction curve is, however, more difficult to obtain, since failure tests involving many combinations of strain cycles and hold times must be performed for each material.

9.3 STRAIN RANGE PARTITIONING

Our next method was first proposed by Manson, Halford, and Hirschberg [6]. It is aimed at separating a stress-strain cycle into its component behaviors and then determining the damage caused by each. This is in contrast to the previous method, in which the total strain range was simply determined without regard to its being tensile or compressive, and without taking cognizance of how much strain was due to creep and how much strain was due to plasticity. Since these factors all influence damage in their own ways, the method of strain range partitioning was offered as an improvement to the method we have just described.

Let us consider a typical stress-strain cycle at a point in a structure as described in Fig. 9.6. This is assumed to be a repetitive cycle and is referred to as a hysteresis loop. The total inelastic range is constant; that is, there is no ratchetting going on. In any such loop there is, in general, tensile plastic strain AC followed by tensile creep CD at constant stress. Then there is compressive plastic strain DB followed by compressive creep

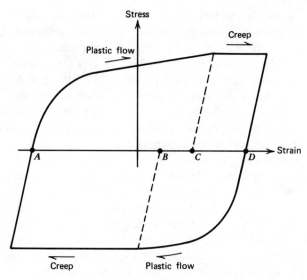

Fig. 9.6 Hysteresis loop [6]. Courtesy of the American Society of Mechanical Engineers.

BA. The two plastic components AC and DB are not necessarily equal in magnitude and the two creep components CD and BA are not necessarily equal. However, since the loop is closed, the tensile strain AD is equal to the compressive strain DA. The total strain range is now partitioned as follows: a completely reversed plastic strain range $\Delta\varepsilon_{pp}$ is defined as the smaller of the two plastic components; in this case it is DB. A completely reversed creep strain range $\Delta\varepsilon_{cc}$ is defined as the smaller of the two creep components, in this case CD. From Fig. 9.6 it can be seen that the difference between the two plastic components is equal to the difference between the two creep components or $AC - DB = BA - CD$. The difference is called $\Delta\varepsilon_{pc}$ if the tensile plastic strain is greater than the compressive plastic strain, as is the case in Fig. 9.6. It is called $\Delta\varepsilon_{cp}$ if the compressive plastic strain is greater than the tensile plastic strain. Now set up the following summary notation:

$\Delta\varepsilon_{pp}$ completely reversed plasticity

$\Delta\varepsilon_{pc}$ tensile plasticity reversed by compressive creep

$\Delta\varepsilon_{cp}$ tensile creep reversed by compressive plasticity

$\Delta\varepsilon_{cc}$ completely reversed creep

In the notation the first subscript (c for creep, p for plasticity) refers to the type of strain imposed in the tensile portion of the cycle, and the second subscript refers to the type of strain imposed in the compressive portion of the cycle. A cycle will contain either $\Delta\varepsilon_{pc}$ or $\Delta\varepsilon_{cp}$, not both. The total inelastic strain range will be equal to the sum of the partitioned strain ranges. Thus we see that since the cycle has been broken down into its component behaviors, it should be possible to assess the contribution to the damage that is caused by each. Before doing this, we should comment on the nature of the various components.

The completely reversed plastic strain range $\Delta\varepsilon_{pp}$ is the basis of low cycle fatigue at temperatures below the creep range of a material. At high temperatures it is the important component only for high frequencies. The mixed strain ranges $\Delta\varepsilon_{pc}$ and $\Delta\varepsilon_{cp}$ are typical in thermal stress problems. The former also occurs when there is elevated temperature strain cycling with compressive hold times, while the latter occurs in this type of cycling with tensile hold times. The completely reversed creep strain range $\Delta\varepsilon_{cc}$ is encountered in cyclic creep rupture tests and low frequency, high temperature reversed loading fatigue tests.

Now let us discuss the damage caused by the various strain ranges and the types of tests that are needed to determine each. The basic assumption is that the damage is described by a Coffin-Manson [3,7,8] type relationship that was first promulgated for completely reversed plastic strain ranges at room temperature. It takes the general form

$$N_f \Delta\varepsilon^\alpha = C \qquad (9.2)$$

Here, each term pertains to the type of strain range that will be considered. N_f is the number of cycles to failure, $\Delta\varepsilon$ is the type of strain range, and α and C are experimentally determined constants.

The methods of determining the life resulting from the completely reversed plastic strain range $\Delta\varepsilon_{pp}$ are well understood at room temperature [3]. At high temperature they are, however, strongly influenced by the frequency of the straining. At low frequencies there is time during the cycle for creep damage to occur, while at high frequencies such damage cannot develop. Thus the life due to $\Delta\varepsilon_{pp}$ must be measured at sufficiently high frequencies so that one can be sure the damage is due only to plastic cycling. The measured cyclic life run at the proper frequency as just described is referred to as N_{pp}. The desired form of the life relationship is then obtained from a straight line fit of a plot of log ($\Delta\varepsilon_{pp}$) versus log (N_{pp}). Typical results for such data using $2\frac{1}{4}$ Cr-1 Mo steel are shown in Fig. 9.7 [6] along with the hysteresis loop of the test. Since no creep takes place, the increment BD is identically equal to $\Delta\varepsilon_{pp}$. A straight line fit to

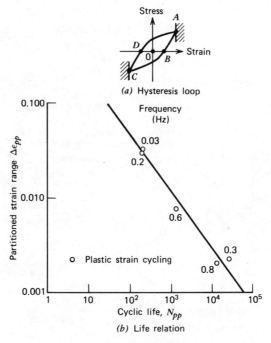

Fig. 9.7 Life relation for completely reversed plastic strain. $2\frac{1}{4}$ Cr-1Mo steel, 1100°F (865 K) [6]. Courtesy of the American Society of Mechanical Engineers.

the curve in Fig. 9.7 gives the relationship*

$$N_{pp}^{0.60}\Delta\varepsilon_{pp}=0.74 \qquad (9.3)$$

There are two main ways to determine the life resulting from $\Delta\varepsilon_{pc}$, the component of tensile plasticity reversed by compressive creep. Both involve constant temperature that is in the creep range. These are shown in Figs. 9.8a and b. In the first, tensile straining is applied at a high rate so that plastic strain without creep strain is produced. The compressive portion of the cycle involves a hold period at constant stress. This permits the creep strain to be easily measured. In Fig. 9.8a the tensile plastic strain is EB, the compressive creep strain is FE, and a small amount of plastic compressive strain is BF. For this cycle, $\Delta\varepsilon_{pc}$ is EF, and $\Delta\varepsilon_{pp}$ is FB. In the second type of test both rapid tensile and compressive straining are

---

*While this is different in form from Eq. (9.2), it is equivalent to it.

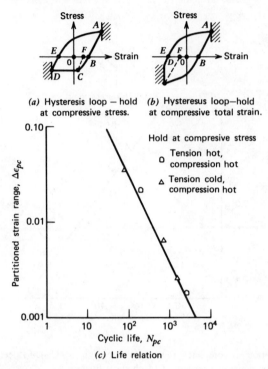

Fig. 9.8 Life relation for tensile plastic strain reversed by compressive creep strain. $2\frac{1}{4}$ Cr-1Mo steel, 1100°F (865 K) and 600°F (590 K) [6]. Courtesy of the American Society of Mechanical Engineers.

imposed but a hold at constant peak compressive strain is maintained to permit compressive stress relaxation. The cycle is shown in Fig. 9.8b has $\Delta\varepsilon_{pc}$ given by EF, and $\Delta\varepsilon_{pp}$ given by FB. Both tests have $\Delta\varepsilon_{pc}$ and $\Delta\varepsilon_{pp}$. For this reason it is best to determine the life caused by $\Delta\varepsilon_{pp}$ first, as described previously. Then, using the linear damage assumption, we can write the life N caused by both as

$$\frac{N}{N_{pp}} + \frac{N}{N_{pc}} = 1 \tag{9.4}$$

In the tests the combined N is measured, as are $\Delta\varepsilon_{pc}$ and $\Delta\varepsilon_{pp}$. Since N_{pp} due to $\Delta\varepsilon_{pp}$ is obtained from Eq. (9.3), we can solve Eq. (9.4) for N_{pc}. Then, a straight line fit of log (N_{pc}) versus log ($\Delta\varepsilon_{pc}$) gives the desired relationship. For the $2\frac{1}{4}$ Cr-1 Mo material, this is shown in Fig. 9.8, and

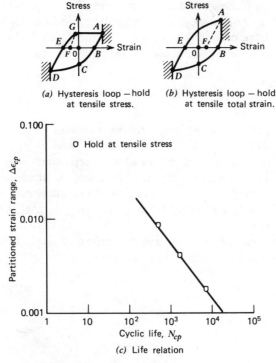

Fig. 9.9 Life relation for tensile creep strain reversed by compressive plastic strain. $2\frac{1}{4}$ Cr-1Mo steel, 1100°F (865 K) [6]. Courtesy of the American Society of Mechanical Engineers.

the equation is [6]

$$N_{pc}^{0.91}\Delta\varepsilon_{pc}=2.15 \tag{9.5}$$

The life caused by $\Delta\varepsilon_{cp}$, the component of tensile creep reversed by compressive plasticity, is obtained in a manner similar to the way that the $\Delta\varepsilon_{pc}$ component was studied. Now the conditions under which the tensile and compressive strains are applied are reversed. The two tests are shown in Fig. 9.9a and b. $\Delta\varepsilon_{cp}$ is FB and $\Delta\varepsilon_{pp}$ is EF in both. Again, a $\Delta\varepsilon_{pp}$ component occurs but, using the linear damage relationship for the combined life N, we write

$$\frac{N}{N_{pp}}+\frac{N}{N_{cp}}=1 \tag{9.6}$$

Having measured N, $\Delta\varepsilon_{cp}$, and $\Delta\varepsilon_{pp}$, and using Eq. (9.3) to determine N_{pp}, we can solve Eq. (9.6) for N_{cp}. Then, a straight line fit of log (N_{cp}) versus log ($\Delta\varepsilon_{cp}$) gives the desired relationship. For the $2\frac{1}{4}$ Cr-1 Mo material this is shown in Fig. 9.9 and the equation is [6]

$$N_{cp}^{0.60}\Delta\varepsilon_{cp}=0.32 \tag{9.7}$$

The last test to be run determines the life resulting from the completely reversed creep component $\Delta\varepsilon_{cc}$. There are two ways of carrying this out. One is the cyclic creep rupture test, in which creep occurs at constant stress at two points. The other is the strain cycling test, in which two relaxation periods at constant strain occur. Examples of this are shown in Fig. 9.10a and b. During this type of test, both $\Delta\varepsilon_{pp}$ and $\Delta\varepsilon_{pc}$ (or $\Delta\varepsilon_{cp}$) ranges occur in addition to $\Delta\varepsilon_{cc}$.

For this reason the present test is performed last. Linear damage summation then is used in the form:

$$\frac{N}{N_{pp}}+\frac{N}{N_{pc}}+\frac{N}{N_{cc}}=1 \tag{9.8a}$$

or

$$\frac{N}{N_{pp}}+\frac{N}{N_{cp}}+\frac{N}{N_{cc}}=1 \tag{9.8b}$$

depending on which is applicable. Now N, $\Delta\varepsilon_{pc}$ (or $\Delta\varepsilon_{cp}$), and $\Delta\varepsilon_{pp}$ have been measured. N_{pp} is obtained with Eq. (9.3) from $\Delta\varepsilon_{pp}$, and N_{pc} (or N_{cp}) is obtained with Eq. (9.5) [or Eq. (9.7)] with $\Delta\varepsilon_{pc}$ (or $\Delta\varepsilon_{cp}$). From these and the appropriate equation of Eqs. (9.8a,b) N_{cc} can then be calculated. Then, a straight line fit of log (N_{cc}) versus log ($\Delta\varepsilon_{cc}$) gives the desired relationship. For the $2\frac{1}{4}$ Cr-Mo material, this is shown in Fig. 9.10 and the equation is [6]

$$N_{cc}^{0.72}\Delta\varepsilon_{cc}=0.76 \tag{9.9}$$

For a general cycle that has been partitioned into $\Delta\varepsilon_{pp}$, $\Delta\varepsilon_{cc}$, and $\Delta\varepsilon_{pc}$ (or $\Delta\varepsilon_{cp}$) contributions, the life N can be found by the linear damage summation assumption by using the form of Eq. (9.8a,b) that applies. Note that in a given cycle either $\Delta\varepsilon_{pc}$ or $\Delta\varepsilon_{cp}$ may occur, but not both. An interactive damage summation rule has also been proposed by Manson [9]. In this method we call $\Delta\varepsilon_i$ the total inelastic strain range in the hysteresis

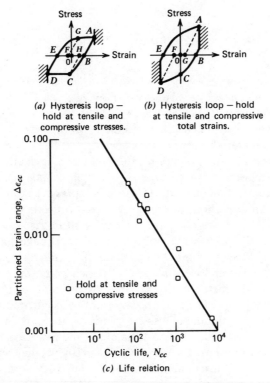

Fig. 9.10 Life relation for completely reversed creep strain. $2\frac{1}{4}$ Cr-1Mo steel, 1100°F (865 K) [6]. Courtesy of the American Society of Mechanical Engineers.

loop; $\Delta\varepsilon_{pp}$, $\Delta\varepsilon_{cc}$, $\Delta\varepsilon_{cp}$ (or $\Delta\varepsilon_{pc}$) are as before. Calculate the fractions

$$F_{pp} = \frac{\Delta\varepsilon_{pp}}{\Delta\varepsilon_i} \qquad F_{cp} = \frac{\Delta\varepsilon_{cp}}{\Delta\varepsilon_i}$$

$$F_{cc} = \frac{\Delta\varepsilon_{cc}}{\Delta\varepsilon_i} \qquad F_{pc} = \frac{\Delta\varepsilon_{pc}}{\Delta\varepsilon_i}$$

(9.10)

The interaction damage rule is then

$$\frac{1}{N_f} = \frac{F_{pp}}{N_{pp}} + \frac{F_{cp}}{N_{cp}} + \frac{F_{cc}}{N_{cc}}$$

or

$$\frac{1}{N_f} = \frac{F_{pp}}{N_{pp}} + \frac{F_{pc}}{N_{pc}} + \frac{F_{cc}}{N_{cc}} \qquad (9.11)$$

depending on which applies. However, now N_{pp}, N_{cp}, N_{pc}, and N_{cc} are obtained from equations such as Eq. (9.3), (9.5), (9.7), or (9.9), using $\Delta\varepsilon_i$ in each case. Recall however that the specific form of these pertains only to the $2\frac{1}{4}$ Cr-1Mo material.

The chief strengths of the strain range partitioning method are that it obtains the damage from each type of strain that may occur in a given cycle. In particular, it includes the effect of compressive strain without any particular assumption about it. However, one still needs a complete stress analysis; in fact, the hysteresis loop at every point in the structure must be obtained so that the partitioning and, thereby, the damage calculation, can be carried out. This is more complicated than the calculation of the effective strain range and effective stress at each point, as is called for by the A.S.M.E. Code method. In fact, the partitioning is not usually as obvious as it is in Fig. 9.6. This is because in the usual situation the creep and plasticity occur simultaneously, so various methods of construction must be used to accomplish the partitioning of the hysteresis loop. Several such methods were proposed by Manson et al. [10] but they are beyond the scope of the present discussion. Moreover, the discussion as presented pertains only to uniaxial situations. For the multiaxial cases usually encountered in design Manson [1] has proposed that the hysteresis loop be determined on the basis of effective strains and stresses. However, this introduces further complications into the design analysis process because the signs of the effective quantities are, by virtue of their definitions, always positive. Thus rules must be set up to specify the signs of the effective stresses and strains. This is most often based on the sign of the predominant stress and is discussed in more detail by Manson in Ref. [1]. Finally, the method requires that four tests be run to characterize a given material. These determine the constants in Eq. (9.2) for each of the four types of strain components that could occur in a given situation.

9.4 FREQUENCY SEPARATION

The Coffin-Manson law Eq. (9.2) governs what is known as low cycle fatigue in which the primary damage is caused by plastic straining. If the straining is elastic, the behavior is known as high cycle fatigue. The transition between the two behaviors is usually around 10^4 cycles. Low cycle fatigue is characterized by wide hysteresis loops, while high cycle fatigue has very narrow hysteresis loops. Elevated temperature fatigue is most often involved with inelastic straining and is, therefore, a low cycle fatigue problem. We have already discussed the effect of frequency on the reversed plastic straining test in the preceding section. There we mentioned that low frequency cycling at elevated temperature permits creep damage to enter, whereas at high frequency such damage is excluded. Coffin [11] has proposed a frequency modified, Coffin-Manson law to account for this as follows. Starting with Eckel's observation [12] that

$$\nu^k t_f = \text{constant} = f(\Delta \varepsilon_p) \qquad (9.12)$$

where ν is the frequency, t_f is the time to failure, and k is a temperature dependent material constant. The term on the left hand side can be rewritten as

$$\nu^k t_f = \nu^k \frac{N_f}{\nu} = N_f \nu^{k-1} \qquad (9.12a)$$

This is known as the frequency modified fatigue life. It is used in the Coffin-Manson equation for elevated temperature, low cycle fatigue, so that this equation becomes

$$\left(N_f \nu^{k-1}\right) \Delta \varepsilon_p^\alpha = C \qquad (9.13)$$

Coffin writes this equation in the equivalent alternate form

$$\left(N_f \nu^{k-1}\right)^\beta \Delta \varepsilon_p = C_1 \qquad (9.13a)$$

Note that when $k=1$, the room temperature equation is obtained and there is no frequency dependence. Coffin [11] has proposed this to correlate elevated temperature fatigue tests. Some of these are shown in Fig. 9.11 for 304 stainless steel. The data were obtained by Berling and Slot [13]. Note that Eq. (9.13a) with $k \neq 1$ pertains to the right of the transition point while the value $k=1$ pertains to the left of the transition point. Note that the air tests show a pronounced frequency effect while the vacuum test does not, even at elevated temperature. This is the basis of Coffin's contention that it is not creep but environment that influences elevated temperature fatigue results [11],[1].

To introduce the influence of the shape of the hysteresis loop into the frequency modified fatigue equations, Coffin proposed [1] the following approach: Fig. 9.12 shows hysteresis loops of stress and inelastic strain

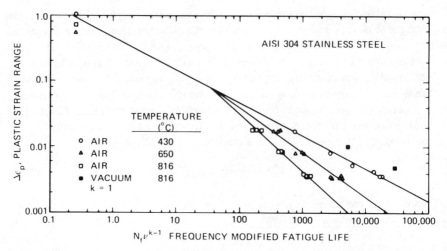

Fig. 9.11 Plastic strain range vs fatigue life of AISI 304 stainless steel in air and vacuum [13]. Courtesy of the Oak Ridge National Laboratory.

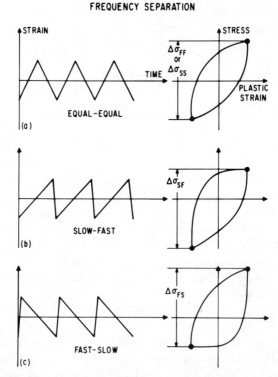

Fig. 9.12 Effect of equal and unequal ramp rates on hysteresis loop shape [1]. Courtesy of the Oak Ridge National Laboratory.

corresponding to equal and unequal plastic *strain rates* (slopes). For equal slopes, as in Fig. 9.12a, the stress range $\Delta\sigma_{SS}$ for slow equal rates is less than $\Delta\sigma_{FF}$ for fast equal rates. When the strain rates are unequal, as in Fig. 9.12b and c, the loops become unbalanced. When the rate is slow the corresponding stress prior to strain reversal is low; when the rate is fast the stress produced is high. Thus a slow-fast loop will be lower in tension and higher in compression, as seen in Fig. 9.12b, while a fast-slow loop is just the opposite, as seen in Fig. 9.12c. The stress ranges for these loops are $\Delta\sigma_{SF}$ and $\Delta\sigma_{FS}$. The relationship between plastic strain range and stress range for a balanced equal strain rate loop has been found to be [1]

$$\Delta\sigma = A(\Delta\varepsilon_p)^{n'} \nu^{k_1} \tag{9.14}$$

where A, n', and k_1 are constants. This is a frequency modified relation. To apply this to an unbalanced loop, say slow-fast, we will have a problem because the two frequencies will not be the same. Thus Eq. (9.14) will predict different stress ranges for the two frequencies (or strain rates). Coffin assumed that the actual stress range was the average of these two results. Thus for the tensile-going part

$$\Delta\sigma_S = A(\Delta\varepsilon_p)^{n'} \left(\frac{\nu_t}{2}\right)^{k_1} \tag{9.14a}$$

and for the compression-going part

$$\Delta\sigma_F = A(\Delta\varepsilon_p)^{n'} \left(\frac{\nu_c}{2}\right)^{k_1} \tag{9.14b}$$

Thus the average is

$$\Delta\sigma_{SF} = \frac{A}{2}(\Delta\varepsilon_p)^{n'} \left(\left(\frac{\nu_c}{2}\right)^{k_1} + \left(\frac{\nu_t}{2}\right)^{k_1} \right) \tag{9.14c}$$

In this approach $\Delta\sigma_{FS} \equiv \Delta\sigma_{SF}$. Now a problem arises because Eq. (9.14) demands a different stress range than is given in Eq. (9.14c), although the plastic strain range $\Delta\varepsilon_p$ is the same in each. This is seen because the average frequency is obtained from the sum of the periods; that is

$$t = t_c + t_t \tag{9.15}$$

or

$$\frac{1}{\nu} = \frac{1}{\nu_c} + \frac{1}{\nu_t}, \qquad \nu = \frac{1}{\nu_t^{-1} + \nu_c^{-1}} \tag{9.15a}$$

and Eq. (9.14) gives

$$\Delta\sigma = A(\Delta\varepsilon_p)^{n'} \left(\frac{1}{v_t^{-1} + v_c^{-1}} \right)^{k_1} \quad (9.15b)$$

Equation (9.15b) is not equal to Eq. (9.14c) unless $v_c = v_t$. To resolve this we assume that there is an equivalent plastic strain range $\Delta\varepsilon_p^1$ associated with the stress range $\Delta\sigma_{FS}$ in Eq. (9.14c) and the tension-going frequency v_t only. Thus Eqs. (9.14) and (9.14c) give

$$\Delta\varepsilon_p^1 = \left[\frac{\Delta\sigma_{SF}}{A} \frac{1}{\left(\frac{v_t}{2} \right)^{k_1}} \right]^{1/n'}$$

$$= \Delta\varepsilon_p \left(\frac{(v_c/v_t)^{k_1} + 1}{2} \right)^{1/n'} \quad (9.16)$$

and then from Eq. (9.13a) the cycles to failure are obtained as

$$N_f = \left(\frac{C_1}{\Delta\varepsilon_p^1} \right)^{1/\beta} \left(\frac{v_t}{2} \right)^{1-k} \quad (9.17)$$

This tells us that according to this approach the damage depends on the tension-going time and on the unbalance in the loop. The use of v_t in Eqs. (9.16) and (9.17) accounts for the former and the use of $\Delta\varepsilon_p^1$ accounts for the latter. If we consider three loops of the same total period we will have $(\Delta\varepsilon_p^1)_{SF} > (\Delta\varepsilon_p^1)_{\text{equal}} > (\Delta\varepsilon_p^1)_{FS}$ and $(N_f)_{SF} < (N_F)_{\text{equal}} < (N_f)_{FS}$. This is in accord with observed facts. Another, more useful form is obtained by combining Eqs. (9.16) and (9.17) to give

$$N_f = \left(\frac{C_1}{\Delta\varepsilon_p} \right) \left(\frac{v_t}{2} \right)^{1-k} \left[\frac{\left(\frac{v_c}{v_t} \right)^{k_1} + 1}{2} \right]^{1/\beta n'} \quad (9.18)$$

We may compare this to a general correlation equation of the form

$$N_f = D(\Delta\varepsilon_p)^a v_t^b \left(\frac{v_c}{v_t} \right)^c \quad (9.19)$$

Eqs. (9.18) and (9.19) are equivalent if

$$\frac{\left(\frac{v_c}{v_t} \right)^{k_1} + 1}{2} \simeq \left(\frac{v_c}{v_t} \right)^{k_1/2} \quad (9.19a)$$

Therefore, an alternative form for the damage is

$$N_f = \left(\frac{C_1}{\Delta \varepsilon_p}\right)^{1/\beta} \left(\frac{\nu_t}{2}\right)^{1-k} \left(\frac{\nu_c}{\nu_t}\right)^c \qquad (9.20)$$

Now, C_1, β and k are found from balanced loop data, and c is found from unbalanced loop data.

Coffin [1] discusses several additional matters concerning the frequency separation approach, namely, the handling of wave shapes that are more complicated than those shown in Fig. 9.12, cumulative damage due to several cycles, and multiaxial situations. Note that no mention is made of creep strain. The development simply assumes that all inelastic strain is included in the term "plastic strain." This is quite different from the intent of strain range partitioning, where the nature of each component of strain is felt to be important. However, the use of the Coffin-Manson formula in both approaches gives them a common phenomenological base. Since our purpose here is to introduce these ideas, we do not delve into them further.

9.5 CLOSURE

In this chapter we endeavored to introduce the reader to the evaluation of stress-strain analysis results with respect to the creep-fatigue interaction problem. We presented several methods of which the A.S.M.E. Code method represents an approach that has been adopted by one industry. There are, however, at least two other methods that have been proposed to evaluate cyclic stress-strain histories. These are the method of ductility exhaustion proposed by Polhemus et al. [14] and the method based on plots of rupture time t_R versus cyclic life N_f that was proposed by Ellis and Esztergar [15]. We have not presented these methods, nor have we gone into greater detail for strain range partitioning and frequency separation, in the interest of brevity and also because the field is still emerging. We should emphasize that, so far, all attempts at confirmation of these methods have been one-dimensional and have involved idealized histories such as the one we presented in Chapter 1. The greatest difficulty in arriving at a method that is suitable for use in evaluating industrial components is that of verification for the multiaxial stress-strain histories that pertain to such structures. These are typically required to operate for as much as twenty years under load conditions that can, at best, be only guessed at the start.

A comparison of two of the methods that have been discussed here was carried out for a multiaxial problem by Lobitz and Nickell [16]. They used the A.S.M.E. Code method and strain range partitioning to determine the

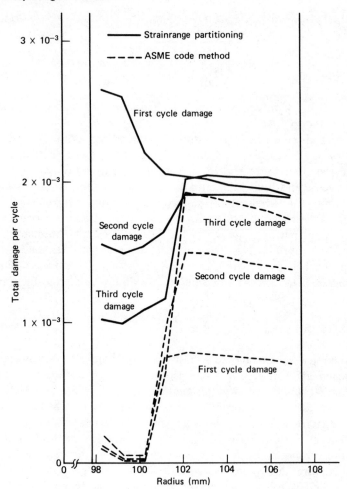

Fig. 9.13 Total damage/cycle versus radius [16].

damage for each of the first three cycles in the pipe ratchetting test that we discussed in Chapters 7 and 8. Their results are shown in Fig. 9.13 where the calculated total damage per cycle is plotted across the wall of the pipe for the three cycles. It is immediately obvious that the two approaches give vastly different estimations of the damage. The A.S.M.E. Code method predicts increasing damage for the outer portion of the tube while strain range partitioning predicts decreasing damage for the inner portion of the tube. Since the test was not continued to failure one cannot determine which approach is the most accurate. Thus it is obvious that one needs to carry out multiaxial tests to failure before these or any of the other methods that have been discussed can be determined to be the best. Such testing is complicated and expensive, and, therefore, not readily undertaken, as is the testing of actual components for verification of these design rules. Thus it is expected that the resolution of the problem that we addressed in this chapter will require a great deal of time, expense, and particularly, operating experience. Therefore, the reader is urged to follow the current literature on this evolving area of effort.

REFERENCES

[1] L. F. Coffin, S. S. Manson, A. E. Carden, L. K. Severud, and W. L. Greenstreet, Time Dependent Fatigue of Structural Alloys, ORNL-5073, Oak Ridge National Laboratory, TN, January 1977.

[2] A.S.M.E. Boiler and Pressure Vessel Code, Case Interpretations, Code Case N-47, American Society of Mechanical Engineers, New York, (1974).

[3] C. C. Osgood, *Fatigue Design*, Wiley-Interscience, New York, (1970).

[4] S. Y. Zamrik, The Complexity of Biaxial Fatigue Analysis in the Low Cycle Region, in S. Y. Zamrik, and R. I. Jetter, editors, *Advances in Design for Elevated Temperature Environment*, A.S.M.E., New York, pp. 39–44 (1975).

[5] M. Kitagawa, and R. W. Weeks, Analysis of Hold Time Fatigue Test Results of AISI 304 Stainless Steel by Five Existing Methods, *Proc. Symp. on Mechanical Behavior of Materials*, Kyoto, Japan, pp. 233–38 (1973).

[6] S. S. Manson, G. R. Halford, and M. H. Hirschberg, Creep Fatigue Analysis by Strain Range Partitioning, in S. Y. Zamrik, editor, *Design for Elevated Temperature Environment*, A.S.M.E., New York, pp. 12–28 (1971).

[7] S. S. Manson, Behavior of Materials Under Conditions of Thermal Stress, NACA-TN-2933 (1953).

[8] L. F. Coffin, Jr., A Study of the Effects of Cyclic Thermal Stresses on a Ductile Material, *Trans. A.S.M.E.*, vol. 76, pp. 931–950 (1954).

[9] S. S. Manson, The Challenge to Unify Treatment of High Temperature Fatigue—A Partisan Proposal Based on Strain Range Partitioning, *Fatigue at Elevated Temperatures*, A.S.T.M. STP 520, American Society for Testing and Materials, pp. 744–782 (1973).

[10] S. S. Manson, G. R. Halford and A. J. Nachtigall, Separation of the Strain Component for Use in Strain Partitioning, in S. Y. Zamrik, and R. I. Jetter, editors, *Advances in Design for Elevated Temperature Environment*, A.S.M.E., New York, pp. 17–28 (1975).

[11] L. F. Coffin, Jr., Fatigue at High Temperature—Predictions and Interpretations, *I. Mech. Engrs. Proceedings*, vol. 188, pp. 109–127 (1974).

[12] J. F. Eckel, The Influence of Frequency on the Repeated Bending Life of Acid Lead, *Proc. A.S.T.M.*, vol. 51, pp. 745–756 (1951).

[13] J. T. Berling, and T. Slot, Effect of Strain Rate on Low Cycle Fatigue Resistance of AISI 304, 316, and 348 Stainless Steels at Elevated Temperature, *Symp. on High Temperature*, A.S.T.M. 465, pp. 3–30 (1968).

[14] J. F. Polhemus, C. E. Spaeth, and W. H. Vogel, Ductility Exhaustion Model for Predicting Thermal Fatigue and Creep Interaction, Fatigue at Elevated Temperatures, A.S.T.M.-STP-520, pp. 625–35 (1972).

[15] J. R. Ellis, and E. P. Esztergar, Considerations of Creep Fatigue Interaction in Design Analysis, in S. Y. Zamrik, editor, *Design for Elevated Temperature Environment*, A.S.M.E., New York, NY, pp. 29–43 (1971).

[16] D. W. Lobitz, and R. E. Nickell, Creep Fatigue Damage Under Multiaxial Conditions, Sandia Laboratories Report SAND 76-0720, Albuquerque, NM, February 1977.

[17] Criteria for Design of Elevated Temperature Class 1 Components in Section III, Division 1, of the A.S.M.E. Boiler and Pressure Vessel Code, A.S.M.E., New York, NY, May 1976.

[18] R. D. Campbell, Creep-Fatigue Calculation Procedures for Code Case 1592, in S. Y. Zamrik, and R. I. Jetter, editors, *Advances in Design for Elevated Temperature Environment*, A.S.M.E., New York, pp. 45–56 (1975).

AUTHOR INDEX

Almroth, B. O., 126, 148
Anderson, R. G., 65, 67, 68, 95
Argon, A. S., 17
Argyris, J. H., 190

Bathe, K. J., 190
Baur, L., 145, 148
Becker, H., 148
Beer, F. J., 151, 176
Berling, J. T., 235, 241
Boardman, F. D., 84, 96
Boley, B. A., 127, 148
Boyle, J. T., 66, 82, 96
Bree, J., 151-3, 155, 167, 176
Brush, D. O., 126, 148
Burgreen, D., 151, 155, 167, 176

Campbell, R. D., 242
Carden, A. E., 219, 241
Carlson, R. L., 148
Charman, C., 216
Chern, J., 82, 96, 136, 139, 141-5, 147-8, 201-2, 216
Clinard, J. A., 198, 205, 216
Coffin, L. F., Jr., 219, 225, 228, 235, 237, 239, 241
Corum, J. M., 172, 176, 198, 200, 203, 215-6

Dhalla, A., 190-1, 207, 212, 216-7
Donnell, L. H., 148
Drucker, D. C., 91, 97

Eckel, J. F., 235, 241
Edamoto, T., 190
Edmunds, H. G., 151, 176
Ellen, F. P., 96
Ellis, J. R., 239, 241
Esztergar, E. P., 239, 241

Fairbairn, J., 84, 89, 96, 122, 124
Finnie, I., 17, 45, 52, 62

Gallagher, R. H., 178, 185, 190-1, 207, 215-6
Gardner, L. R. T., 95
Garofalo, F., 17
Gatewood, B. E., 151, 176
Gerard, G., 126, 131, 148
Gerdeen, J. C., 148
Gere, J. M., 126, 148
Goel, R. P., 121, 124
Goodier, J. N., 62
Greenbaum, G. A., 192, 216
Greenstreet, W. L., 219, 241
Grindell, A. G., 172, 176, 216

Haferd, A. M., 103-4, 123
Haisler, W. E., 189, 216
Halford, G. R., 226, 241
Hayhurst, D. R., 85, 87, 90, 97, 118-22, 124
Hellen, T. K., 190
Heller, W. R., 17, 45, 52, 62
Henrywood, R. K., 190

Author Index

Henschell, R. D., 190
Hildebrand, F. B., 216
Hirschberg, M. H., 186, 216, 226, 241
Hodge, P. G., 72, 96
Hodgkins, W. R., 95
Hoff, N. J., 44, 110, 127, 129, 148
Hult, J. A. H., 17, 45, 62, 79-81, 86, 96-7

Ishii, A., 190

Jetter, R. I., 241-2
Johnson, W., 17, 45
Johnsson, A., 66, 68, 77, 95
Juvinall, R. C., 17

Keller, H. B., 148
Kelly, D. A., 97, 124
Kitagawa, M., 225, 241
Koiter, W. T., 97
Kraus, H., 97, 202, 215-7
Krempl, E., 19, 44

Larson, F. R., 100-2, 104, 123
Leckie, F. A., 72, 82, 85, 87, 89, 95-7, 118-9, 121, 124
Lobitz, D., 239, 241

McClintock, F. A., 17
Mackenzie, A. C., 67-8, 72, 80, 82, 96
Manjoine, M. J., 216
Manson, S. S., 17, 103-4, 123-4, 186, 216, 219, 226, 228, 234, 241
Marriott, D. L., 17, 21-2, 45, 67, 82, 96, 118, 122-4, 134-5, 145-6, 148
Martin, J. B., 92-3, 97, 121, 124
Mellor, P. B., 17, 45
Mendelson, A., 34, 44, 186, 195, 216
Miller, D. R., 151, 167, 176
Miller, J., 100-2, 104, 123
Minato, A., 190
Mori, T., 190
Morrison, C. J., 85, 97, 124
Murakami, T., 190

Nachtigall, A. J., 241
Nadai, A., 138, 148
Nagato, K., 190, 216
Nickell, R. E., 190, 206-7, 216-7, 239, 241

O'Donnell, W. J., 170, 172, 176
Odqvist, F. K. G., 17, 45, 114, 117, 124, 127, 148
Osgood, C. C., 241

Pai, D. H., 201-2, 216
Palmer, A. C., 95, 97
Penny, R. K., 17, 21-2, 45, 67, 69, 76, 83, 96, 118, 122-4, 134-5, 145-6, 148
Polhemus, J. F., 239, 241
Ponter, A. R. S., 72, 85, 89, 90, 95-7, 122, 124
Porowski, J., 170, 172, 176
Pugh, C. E., 44, 189, 216

Reiss, E. L., 148
Richard, A. P., 138, 148
Richardson, M., 203, 216
Rubinstein, M. F., 192, 216

Samuelson, L. A., 143-5, 148, 205, 216
Sanders, D. R., 189, 216
Sankey, G. A., 216
Sartory, W. K., 216
Sazawal, V. K., 148
Schulte, C. A., 65, 82, 95
Severud, L. K., 219, 241
Shanley, F. R., 130, 148
Shoemaker, E., 216
Sim, R. G., 68-70, 72, 76-9, 83, 87-9, 96
Slot, T., 235, 241
Soderberg, C. R., 64-5, 95
Sokolnikoff, I. S., 97
Spaeth, C. E., 241
Spence, J., 79-82, 96
Stone, C. M., 206-7, 216
Stowell, E. Z., 138, 148
Suh, N. P., 17
Swanson, J. A., 190

Tanikawa, M., 190
Timoshenko, S. P., 62, 126, 148
Turner, A. P. L., 17
Tzung, F. K., 216

Ueda, M., 190
Ural, O., 178, 185, 216

Vogel, W. H., 241

Wahl, A. M., 195, 216
Wan, C. C., 148
Weeks, R. W., 225, 241
Weiner, J. H., 127, 148
Williams, J. J., 97, 124
Williamson, J. A., 96

Yamada, Y., 216

Yates, D. N., 190
Yokouchi, Y., 190
Young, H. C., 172, 176, 216

Zamrik, S., 225, 241-2
Zienkiewicz, O. C., 178, 185, 189-90, 215

SUBJECT INDEX

ADINA Program, 190-1
American Society of Mechanical Engineers, Boiler and Pressure Vessel Code:
  creep-fatigue interaction, 219-26, 234-40
  creep ratchetting, 176
ANSYS Program, 190-1
Approximate analytical techniques, 64-95
  bounding techniques based on virtual work, 90-95
  reference stress method, 64-90
ASAS Program, 190
ASKA Program, 190
ASTUC Program, 190

Bailey-Norton Law, 20, 24
Beam:
  cantilever, 67, 73, 94, 117
  cyclic loading, 89
  pure bending, 66, 71, 73, 79, 82
  relaxation, 79-82
BERSAFE Program, 190
Bounding techniques:
  strains due to creep ratchetting of tube, 170-2
  virtual work, 90-95
Bree diagram, 167
Bree problem, 151
Brittle behavior, 5
Buckling, 16, 126-147
  axially compressed cylindrical shell, 141-5, 205-7
  column, 127-130

comparison of methods, 141-7
constant critical strain method, 139-40
critical effective strain method, 136-47
isochronous stress-strain method, 140-1
reference stress method, 134-6
secant modulus method, 137-9
shell structures, 134-47
tangent modulus method, 130-3

Circular plate, clamped, 73
  prescribed deflections, 203-5
  supported simply, 73
Coffin-Manson Law, 228, 234-5, 239
Column, creep buckling, 127-130
Computer programs for creep analysis, 189-92
  comparison of capabilities, 191
  illustrative solutions, 192-215
  list of, 190
Constant critical strain method, 139-40
Continuity, 119
Creep, buckling, 16, 126-147
  curve, 6, 7, 11, 18
  effect of stress, 7
  effect of temperature, 8
  primary, 6, 20
  ratchetting, 16, 150-76
  recovery, 10
  rupture, 6, 16, 20, 98-123
  secondary, 6, 20
  strain, 10, 11
  tertiary, 6

Creep-fatigue interaction, 218-41
  ASME code procedure, 219-26
  comparison of methods, 239-40
  ductility exhaustion, 239
  frequency separation, 234-9
  rupture time versus cyclic life, 239
  strain range partitioning, 226-34
Critical effective strain method, 136-47
Cyclic loading, continuous, 11, 13
  hold times with creep, 11, 12, 15
  hold times with relaxation, 11, 12, 14
  power plant component, 211-2
  ratchetting, 16, 150-76
Cylindrical shell, buckling:
  under axial load, 141-5, 205-7
  under external pressure, 141
Cylindrical tube, thick, pressurized:
  combined loadings, 76
  creep deformation, 49-52
  elastic solution, 49
  heated, 78
  radial displacement, 74
  stationary solution, 49-52
Cylindrical tube, thin, pressurized
  alternating plasticity, 158-61
  bounding technique for creep strains, 170-2
  creep deformation, 47-8
  cyclic growth, 165-6, 169
  cyclic thermal stress, 150-76
  damage comparison, 239-41
  experimental observations, 172-4
  interaction diagram, 166
  ratchetting, 161-7
  rupture, 114-7
  shakedown, 155-8
  stress relaxation, 167-70

Damage:
  fatigue, 222-3
  linear assumption, 107, 230-4
  rupture, 104-8, 119-20, 223-4
Ductile behavior, 4, 7
Ductility exhaustion, 239

Elastic analogy, 36
Elasticity, 3
Endurance limit, 11
Environmental attack, 16-7

Equation of state, 19, 20
Equilibrium equation, 34
Euler load, 132
Excessive deformation, 16

Failure modes, 16
Fatigue, 16
  curve, 12
  elevated temperature, 11, 218-41
  multiple stress levels, 107
  room temperature, 11
FESS Program, 190
Finite element method, 178-86
  application to creep, 185-7
  displacement vector, 178
  elasticity matrix, 181
  position matrix, 180
  stiffness matrix, 183
  strain matrix, 180
  virtual work principle, 182-3
Frequency separation, 234-9

Heat conduction equation, 34
Hemispherical shell, buckling, 134-6, 142, 145-6
Hysteresis loop, 226, 234-5

Initial strain method, 186-7, 196-7
Instantaneous behavior, 3, 16
Interaction:
  creep-fatigue, 11, 218-41
  diagram, thin tube, 166, 171
Isochronous stress-strain curves, 11, 132
Isochronous stress-strain method, 139-40

Larson-Miller parameter, 100-2

Manson-Haferd parameter, 103-4
MARC Program, 190-1, 206, 208, 212
Mean ratio, 144
Mechanical behavior of metals, 1-17
Memory Theory, 19
MINAT-X Program, 190
Miner's rule, 221
Modelling of creep, 18
  multiaxial, 27-35
  steady creep approximation, 36-9
  uniaxial, 18-27
Multiaxial creep models, 27-35

Subject Index

stress reversals, 39-44

NEPSAP Program, 190-1
Newton's Method, 188
Nonstationary state, 61
  two bar structure, 55-6
Numerical analysis, 23, 69, 177-215
  application to creep, 185-7
  computer programs for creep analysis, 189-92
  creep buckling of a cylindrical shell, 205-7
  creep ratchetting of a tube, 198-201
  creep and relaxation of a circular plate, 203-5
  creep of a rotating gas turbine seal, 207-10
  creep of a straight tube under bending, pressure and thermal loads, 201-3
  cyclic creep of a seal for a nuclear power plant component, 211-2
  finite element method, 178-85
  illustrative solutions, 192-215
  stress redistribution in pressure vessel and heads, 192-5
  stress redistribution in a rotating disc, 195-8
  time extraction for strain hardening, 187-9

PAFEC Program, 190
PCRAP-2 Program, 190
PLACRE Program, 200, 205
PLANS Program, 191
Plasticity, 3
  flow rules, 33
  yield condition, 33
Portal frame, 85

Ratchetting, 16, 150-176
  bounding technique for strains, 170-2
  experimental observations of, 172-4
Recovery, 10, 46
  in a two bar structure, 59-61
Redistribution, 46, 56, 61
  pressure vessels and heads, 192-5
  rotating disc, 195-8
  see also Stationary state
Reference stress method, 64-90
  buckling, 134-6
  combined loadings, 75-6
  creep under steady loads, 65-75

experimental verification, 82-6
rupture, 118-23
stress relaxation, 79-82
thermal loadings, 77-9
variable loads, 86-90
Reference temperature, 77
Relaxation, 11, 16, 35, 36
  cyclic, 11, 12, 14
  effect on ratchetting of a tube, 167-70
  reference stress method, 79-82
  uniaxial bar, 57-9
Robinson's Rule, 221
Rupture, 6, 16, 20, 98
  analysis, 104-18
  beam in pure bending, 117-8
  damage, 104-6
  data, 99
  Larson-Miller parameter, 100-2
  Manson-Haferd parameter, 103-4
  multiaxial states, 114
  multiple stress levels, 106-8
  nonhomogeneous states, 117
  reference stress method, 118-23
  tensile specimen, 109-14
  thin tube, pressurized, 114-7
  time, 110, 112-4, 116-9, 121-2

SATEPIC Program, 190
Secant modulus method, 137-9
Shell buckling, 134-47
Skeletal point, 82
Spherical shell, buckling, 142
Spherical shell with nozzle, pressurized:
  creep displacement, 86-7
  creep rupture, 122-3
Stability, *see* Buckling
Stationary state, 46, 61
  buckling of a column, 128
  pressurized thick tube, 49-52
  pressurized thin tube, 48
  rectangular beam, 66, 71, 73
  two bar structure, 54-5
  uniformly loaded cantilever, 67, 73
Steady creep approximation, 36-9, 61
Strain:
  conventional, 2, 7
  creep, 19, 33
  deviator, 28, 35
  due to creep ratchetting, 170-2

effective, 30
elastic, 4, 10, 33
engineering, 2
fracture, 4
logarithmic, 3
natural, 2, 7
plastic, 4, 10, 33
rate, 5
rate tensor, 30
total, 33
Strain-displacement equation, 34
Strain hardening formulation, 20, 22-3, 31
  based on primary creep strain, 23
  based on total strain, 23
  pressurized thin tube, 47
  time extraction for, 187-9
Strain range partitioning, 226-34, 39
Stress:
  deviator, 28-9, 35
  effective, 29
  effect on creep, 7, 8
  fracture, 6
  hydrostatic, 28
  nominal, 2, 7
  reference, 64, 90
  relaxation, 11, 46
  reversals, 24
  tensor, 28
  true, 2, 7
  ultimate, 3
  variable, 20
  yield, 3, 6

Tangent modulus method, 130-3

Temperature:
  effect on creep, 7, 9
  reference, 77
  transition, 6
Tensile test, 2, 16
  analysis of creep rupture in, 109-14
Thermal loadings, reference stress method, 77
Thermal strain method, *see* Initial strain method
Time:
  buckling of a column, 130, 133
  buckling of a shell, 137
  rupture, 110, 112-4, 116-9, 121-2
Time dependent behavior, 6, 16
Time extraction for strain hardening creep, 187-9
Time hardening formulation, 20-2, 31
Two bar structure, 52-6
  elastic solution, 53-4
  nonstationary solution, 55-6
  recovery in, 59-61
  stationary solution, 54-5

Uniaxial creep models, 18
  steady creep approximation, 36-7
  stress reversals, 24-7

Variable loadings, reference stress method, 86-90
Virtual work, techniques, 90-5
  in finite element method, 182-3
Volume constancy in creep, 3, 27, 30

WECAN Program, 190-1
Work hardening, 4